One-Factorizations

T0214346

Mathematics and Its Applications

Managing Editor:

M. HAZEWINKEL

Centre for Mathematics and Computer Science, Amsterdam, The Netherlands

Volume 390

One-Factorizations

by

W. D. Wallis
Southern Illinois University,
Carbondale, Illinois, U.S.A.

KLUWER ACADEMIC PUBLISHERS
DORDRECHT / BOSTON / LONDON

Library of Congress Cataloging-in-Publication Data

ISBN 978-1-4419-4766-6

Published by Kluwer Academic Publishers,
P.O. Box 17, 3300 AA Dordrecht, The Netherlands.

Kluwer Academic Publishers incorporates
the publishing programmes of
D. Reidel, Martinus Nijhoff, Dr W. Junk and MTP Press.

Sold and distributed in the U.S.A. and Canada
by Kluwer Academic Publishers,
101 Philip Drive, Norwell, MA 02061, U.S.A.

In all other countries, sold and distributed
by Kluwer Academic Publishers Group,
P.O. Box 322, 3300 AH Dordrecht, The Netherlands.

Printed on acid-free paper

CONTENTS

LIST OF FIGURES

Chapter 23

LIST OF TABLES

PREFACE

This book has grown out of graduate courses given by the author at Southern Illinois University, Carbondale, as well as a series of seminars delivered at Curtin University of Technology, Western Australia. The book is intended to be used both as a textbook at the graduate level and also as a professional reference.

The topic of one-factorizations fits into the theory of combinatorial designs just as much as it does into graph theory. Factors and factorizations occur as building blocks in the theory of designs in a number of places. Our approach owes as much to design theory as it does to graph theory.

It is expected that nearly all readers will have some background in the theory of graphs, such as an advanced undergraduate course in Graph Theory or Applied Graph Theory. However, the book is self-contained, and the first two chapters are a thumbnail sketch of basic graph theory. Many readers will merely skim these chapters, observing our notational conventions along the way. (These introductory chapters could, in fact, enable some instructors to use the book for a somewhat eccentric introduction to graph theory.)

Chapter 3 introduces one-factors and one-factorizations. The next two chapters outline two major application areas: combinatorial arrays and tournaments. These two related areas have provided the impetus for a good deal of study of one-factorizations.

Chapters 6 through 8 are more graph-theoretic. We discuss Tutte's characterization of graphs without one-factors and some applications, and explore a little of the related study of edge-colorings.

Chapters 9 and 10 return to design theory. The relationship to triple systems is another topic which has stirred up some recent interest in one-factorizations. Starters were first defined and investigated as tools in the construction of Room squares and related arrays, and only later were they used primarily to study factorizations.

The next several chapters discuss automorphisms and the variability of one-factorizations. We study ways of recognizing essentially different one-factorizations and give some proofs of the fact that the number of inequivalent one-factorizations of a complete graph goes to infinity with the order of the graph. As an aside, we introduce systems of distinct representatives. Although we use their properties, some readers may wish to skip over this chapter, noting only the results. Chapters 15 through 17 deal with special types of one-factorizations: cyclic, perfect, indecomposable and simple (the latter kinds only arise when considering multigraphs). We then discuss the one-factorization conjecture and related ideas, and close with a study of one-factorizations of graph products.

We have not produced an encyclopedia. The subject of one-factorizations is an amorphous one, and it would be possible to write an arbitrarily large book. Fortunately some related topics have been well covered: there are excellent books on matching theory by Lovasz and Plummer [110], on graph decompositions by Bosák [25] and on edge-colorings by Fiorini and Wilson [66] and Yap [175].

The book contains a number of easy exercises, which should enable the reader to determine whether he or she understands the main ideas. Some harder exercises (marked *) and unsolved problems (marked ?) are also included. The reader of a book at this level should also be involved in reading the current literature: the true exercises, which consist of inventing new directions and proving new results, must come from there.

Several mathematicians have read various drafts and made helpful suggestions. Special thanks in this regard are due to Liz Billington, Dawit Haile, Alex Rosa and Doug West.

1

GRAPHS

Any reader of this book will have some acquaintance with graph theory. However it seems advisable to have an introductory chapter, not only for complete-
ness, but also because writers in this area differ on fundamental definitions: it
is necessary to establish our version of the terminology.

A *graph G* consists of a finite set $V(G)$ of objects called *vertices* together with a
set $E(G)$ of unordered pairs of vertices; the elements of $E(G)$ are called *edges*.
We write $v(G)$ and $e(G)$ for the orders of $V(G)$ and $E(G)$, respectively. In
terms of the more general definitions sometimes used, we can say that "our
graphs are finite and contain neither loops nor multiple edges".

The edge containing x and y is written xy or (x, y); x and y are called its *endpoints*. We say this edge *joins x to y*. $G - xy$ denotes the result of deleting edge
xy from G; if x and y were not adjacent then $G + xy$ is the graph constructed
from G by adjoining an edge xy. Similarly $G - x$ is the graph derived from G
by deleting one vertex x (and all the edges on which x lies). Similarly, $G - S$
denotes the result of deleting some set S of vertices.

A *multigraph* is defined in the same way as a graph except that there may be
more than one edge corresponding to the same unordered pair of vertices. The
underlying graph of a multigraph is formed by replacing all edges corresponding
to the unordered pair $\{x, y\}$ by a single edge xy. Unless otherwise mentioned,
all definitions pertaininging to graphs will be applied to multigraphs in the
obvious way.

If vertices x and y are endpoints of one edge in a graph or multigraph, then
x and y are said to be *adjacent* to each other, and it is often convenient to

write $x \sim y$. The set of all vertices adjacent to x is called the *neighborhood* of x, and denoted $N(x)$. We define the *degree* or *valency* $d(x)$ of the vertex x to be the number of edges which have x as an endpoint. If $d(x) = 0$, x is an *isolated* vertex. A graph is called *regular* if all its vertices have the same degree; in particular, if the common degree is 3, the graph is called *cubic*. We write $\delta(G)$ for the smallest of all degrees of vertices of G, and $\Delta(G)$ for the largest. (One also writes $\Delta(G)$ for the common degree of a regular graph G.) If G has v vertices, so that its vertex-set is, say,

$$V(G) = \{x_1, x_2, \cdots, x_v\},$$

then its *adjacency matrix* M_G is the $v \times v$ matrix with entries m_{ij}, such that

$$m_{ij} = \begin{cases} 1 & \text{if } x_i \sim x_j, \\ 0 & \text{otherwise.} \end{cases}$$

Some authors define the adjacency matrix of a multigraph to be the incidence matrix of the underlying graph; others set m_{ij} equal to the number of edges joining x_i to x_j. We shall not need to use adjacency matrices of multigraphs in this book.

A vertex and an edge are called *incident* if the vertex is an endpoint of the edge, and two edges are called incident if they have a common endpoint. A set of edges is called *independent* if no two of its members are incident, while a set of vertices is independent if no two of its members are adjacent.

If the edge-set is

$$E(G) = \{a_1, a_2, \cdots, a_e\},$$

then the *incidence matrix* N_G of G is the $v \times e$ matrix with entries n_{ij}, such that

$$n_{ij} = \begin{cases} 1 & \text{if vertex } x_i \text{ is incident with edge } a_j, \\ 0 & \text{otherwise.} \end{cases}$$

(The adjacency and incidence matrices depend on the orderings chosen for $V(G)$ and $E(G)$; they are not unique, but vary only by row and column permutation.) The degree $d(x)$ of x will equal the sum of the entries in the row of M_G or of N_G corresponding to x.

Theorem 1.1 *In any graph or multigraph, the number of edges equals half the sum of the degrees of the vertices.*

Proof. It is convenient to work with the incidence matrix: we sum its entries. The sum of the entries in row i is just $d(x_i)$; the sum of the degrees is then

$\sum_{i=1}^{v} d(x_i)$, which equals the sum of the entries in N. The sum of the entries in column j is 2 since each edge is incident with two vertices; the sum over all columns is thus $2e$, so that

$$\sum_{i=1}^{v} d(x_i) = 2e,$$

giving the result. □

Corollary 1.1.1 *In any graph or multigraph, the number of vertices of odd degree is even. In particular, a regular graph of odd degree has an even number of vertices.* □

Given a set S of v vertices, the graph formed by joining all pairs of members of S is called the *complete* graph on S, and denoted K_S. We also write K_v to mean any complete graph with v vertices. The set of all edges of $K_{V(G)}$ which are *not* in a graph G will form a graph with $V(G)$ as vertex-set; this new graph is called the *complement* of G, and written $\overline{G}$. More generally, if G is a subgraph of H, then the graph formed by deleting all edges of G from H is called the *complement of G in H*, denoted $H - G$. The complement $\overline{K}_S$ of the complete graph K_S on vertex-set S is called a null graph; we also write $\overline{K}_v$ for a null graph with v vertices.

An *isomorphism* of a graph G onto a graph H is a one-to-one map ϕ from $V(G)$ onto $V(H)$ with the property that a and b are adjacent vertices in G if and only if $a\phi$ and $b\phi$ are adjacent vertices in H; G is isomorphic to H if and only if there is an isomorphism of G onto H. From this definition it follows that all complete graphs on n vertices are isomorphic. The notation K_n can be interpreted as being a generic name for the typical representative of the isomorphism-class of all n-vertex complete graphs.

If G is a graph, it is possible to choose some of the vertices and some of the edges of G in such a way that these vertices and edges again form a graph, H say. H is then called a *subgraph* of G; one writes $H \leq G$. Clearly every graph G has itself and the 1-vertex graph (which we shall denote K_1) as subgraphs; we say H is a *proper* subgraph of G if it neither equals G nor K_1. If U is any set of vertices of G, then the subgraph consisting of U and all the edges of G which joined two vertices of U is called an *induced* subgraph, the *subgraph induced by U*, and is denoted $\langle U \rangle$. A subgraph G of a graph H is called a *spanning* subgraph if $V(G) = V(H)$. Clearly any graph G is a spanning subgraph of $K_{V(G)}$.

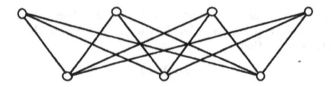

Figure 1.1 $K_{4,3}$.

A graph is called *disconnected* if its vertex-set can be partitioned into two
subsets, V_1 and V_2, which have no common element, in such a way that there
is no edge with one endpoint in V_1 and the other in V_2; if a graph is not
disconnected then it is *connected*. A disconnected graph consists of a number
of disjoint subgraphs; a maximal connected subgraph is called a *component*.

Among connected graphs, some are connected so slightly that removal of a
single vertex of edge will disconnect them. Such vertices and edges are quite
important. A vertex x is called a *cutpoint* in G if $G - x$ contains more components than G does; in particular if G is connected then a cutpoint is a vertex
x such that $G - x$ is disconnected. Similarly a *bridge* (or *cut-edge*) is an edge
whose deletion increases the number of components.

A collection of edges whose deletion disconnects G is called a *cut* in G. A cut
partitions the vertex-set $V(G)$ into two components, A and B say, such that
the edges joining vertices in A to vertices in B are precisely the edges of the
cut, and we refer to "the cut (A, B)". (The two sets A and B are not uniquely
defined — for example, if there is an isolated vertex in G, it could be allocated
to either set — but the cut will be well-defined.) The cut $(A, \{x\})$, consisting
of all edges incident with the vertex x, is called a *trivial* cut.

Generalizing the idea of a cutpoint, we define the *connectivity* $\kappa(G)$ of a graph
G to be the smallest number of vertices whose removal from G results in either
a disconnected graph or a single vertex. (The latter special case is included to
avoid problems when discussing the complete graphs.) If $\kappa(G) \geq k$, then G is
called k-connected. The *edge-connectivity* $\kappa'(G)$ is defined to be the minimum
number of edges whose removal disconnects G (no special case is needed). In
other words, the edge-connectivity of G equals the size of the smallest cut in
G. It is easy to prove (see Exercise 1.6) that

$$\kappa(G) \leq \kappa'(G) \leq \delta(G).$$

The *complete bipartite graph* on V_1 and V_2 has two disjoint sets of vertices, V_1 and V_2; two vertices are adjacent if and only if they lie in different sets. We write $K_{m,n}$ to mean a complete bipartite graph with m vertices in one set and n in the other. Figure 1.1 shows $K_{4,3}$; $K_{1,n}$ in particular is called an *n-star*. Any subgraph of a complete bipartite graph is called "bipartite". More generally, the *complete r-partite graph* $K_{n_1,n_2,\cdots,n_r}$ is a graph with vertex-set $V_1 \cup V_2 \cup \cdots \cup V_r$, where the V_i are disjoint sets and V_i has order n_i, in which xy is an edge if and only if x and y are in different sets. Any subgraph of this graph is called an *r-partite* graph. If $n_1 = n_2 = \cdots = n_r = n$ we use the abbreviation $K_n^{(r)}$.

Several ways of combining two graphs have been studied. The *union* $G \cup H$ and *intersection* $G \cap H$ of graphs G and H are defined in the obvious way — one takes the unions, or intersections, of the vertex-sets and edge-sets (but $G \cap H$ is only defined when G and H have a common vertex). If G and H are edge-disjoint graphs on the same vertex-set, then their union is often also called their *sum* and written $G \oplus H$. At the other extreme, disjoint unions can be discussed, and the union of n disjoint graphs all isomorphic to G is denoted nG.

The notation $G + H$ denotes the *join* of G and H, a graph obtained from G and H by joining every vertex of G to every vertex of H. In these terms

$$K_{m,n} = \overline{K}_m + \overline{K}_n.$$

(No confusion will arise with the earlier use of the $+$ symbol.)

Exercises 1

1.1 Prove that if G is not a connected graph then $\overline{G}$ is connected.

1.2 Suppose G has n vertices and $\delta(G) \geq \frac{n-1}{2}$. Prove that G is connected.

1.3 A graph G is *self-complementary* if G and $\overline{G}$ are isomorphic. Prove that the number of vertices in a self-complementary graph is congruent to 0 or 1 (mod 4).

1.4 Prove that $\overline{G - x} = \overline{G} - x$ for any graph G.

1.5 What is the maximum number of bridges in a graph on v vertices?

1.6 Prove the formula $\kappa(G) \le \kappa'(G) \le \delta(G)$.

1.7 Prove that a graph in which every vertex has even degree can have no bridge.

1.8 G is a bipartite graph with v vertices. Prove that G has at most $\frac{v^2}{4}$ edges.

1.9 Prove that a regular graph of odd degree can have no component with an odd number of vertices.

2

WALKS, PATHS AND CYCLES

A *walk* in a graph G is a finite sequence of vertices $x_0, x_1, \cdots, x_n$ and edges $a_1, a_2, \cdots, a_n$ of G:

$$x_0, a_1, x_1, a_2, \cdots, a_n, x_n,$$

where the endpoints of a_i are x_{i-1} and x_i for each i. A *simple walk* is a walk in which no edge is repeated. A *path* is a walk in which no vertex is repeated; the *length* of a path is its number of edges. A walk is *closed* when the first and last vertices, x_0 and x_n, are equal. A *cycle* of length n is a closed simple walk of length n, $n \geq 3$, in which the vertices $x_0, x_1, \cdots, x_{n-1}$ are all different.

The following observation, although very easy to prove, will be useful.

Theorem 2.1 *If there is a walk from vertex y to vertex z in the graph G, where y is not equal to z, then there is a walk in G with first vertex y and last vertex z.* □

We say that two vertices are *connected* when there is a walk joining them. (Theorem 2.1 tells us we can replace the word "walk" by "path".) Two vertices of G are connected if and only if they lie in the same component of G; G is a connected graph if and only if all pairs of its vertices are connected. If vertices x and y are connected, then their *distance* is the length of the shortest path joining them.

Cycles give the following useful characterization of bipartite graphs.

Theorem 2.2 *A graph is bipartite if and only if it contains no cycle of odd length.*

7

Proof. (i) Suppose G is a bipartite graph with disjoint vertex-sets U and V. Suppose G contains a cycle of odd length, with $2k + 1$ vertices

$$x_1, x_2, x_3, \cdots, x_{2k+1}, x_1$$

where x_i is adjacent to x_{i+1} for $i = 1, 2, \cdots, 2k$, and x_{2k+1} is adjacent to x_1. Suppose x_i belongs to U. Then x_{i+1} must be in V, as otherwise we would have two adjacent vertices in U; and conversely. So, if we assume $x_1 \in U$ we get, successively, $x_2 \in V, x_3 \in U, \cdots, x_{2k+1} \in U$. Now $x_{2k+1} \in U$ implies $x_1 \in V$, which contradicts the disjointness of U and V.

(ii) Suppose that G is a graph with no cycle of odd length. Without loss of generality we need only consider the case where G is connected. Choose an arbitrary vertex x in G, and partition the vertex set by defining Y to be the set of vertices whose distance from x is even, and Z to be the set of vertices whose distance from x is odd; x itself belongs to Y.

Now select two vertices y_1 and y_2 in Y. Let P be a shortest path from x to y_1 and Q a shortest path from x to y_2. Denote by u the last vertex common to P and Q. Since P and Q are shortest paths, so are their sections from x to u, which therefore have the same length. Since the lengths of both P and Q are even, the lengths of their sections from u to y_1 and u to y_2 respectively have equal parity, so the path from y_1 to u (in reverse direction along P) to y_2 (along Q) has even length.

If $y_1 \sim y_2$, then this path together with the edge $y_1 y_2$ gives a cycle of odd length, which is a contradiction. Hence no two vertices in Y are adjacent. Similarly no two vertices in Z are adjacent, and G is bipartite. $\square$

A graph which contains no cycles at all is called *acyclic*; a connected acyclic graph is called a *tree*. Clearly all trees are bipartite graphs.

It is clear that the set of vertices and edges which constitute a path in a graph is itself a graph. We define a walk P_n to be a graph with n vertices $x_1, x_2, \cdots, x_n$ and $n - 1$ edges $x_1 x_2, x_2 x_3, \cdots, x_{n-1} x_n$. A cycle C_n is defined similarly, except that the edge $x_n x_1$ is also included, and (to avoid the triviality of allowing K_2 to be defined as a cycle) n must be at least 3. The latter convention ensures that every C_n has n edges. Figure 2.1 shows P_4 and C_5.

As an extension of the idea of a proper subgraph, we shall define a *proper tree* to be a tree other than K_1, and similarly define a *proper path*. (No definition of a "proper cycle" is necessary.)

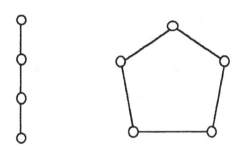

Figure 2.1 P_4 and C_5.

A cycle which passes through every vertex in a graph is called a *Hamilton cycle* and a graph with such a cycle is called *Hamiltonian*. Typically one thinks of a Hamiltonian graph as a cycle with a number of other edges (called *chords* of the cycle). The idea of such a spanning cycle was simultaneously developed by Hamilton [82] in the special case of the icosahédron, and more generally by Kirkman [100].

It is easy to discuss Hamiltonicity in particular cases, and there are a number of small theorems. However, no good necessary and sufficient conditions are known for the existence of Hamilton cycles. The following result is a useful sufficient condition.

Theorem 2.3 *If G is a graph with n vertices, $n \geq 3$, and $d(x) + d(y) \geq n$ whenever x and y are nonadjacent vertices of G, then G is Hamiltonian.*

Proof. Suppose the Theorem is false. Choose an n such that there is an n-vertex counterexample, and select a graph G on n vertices which has the maximum number of edges among counterexamples. Choose two nonadjacent vertices v and w: clearly $G + vw$ is Hamiltonian, and vw must be an edge in every Hamilton cycle. By hypothesis, $d(v) + d(w) \geq n$.

Consider any Hamilton cycle in $G + vw$:

$$v, x_1, x_2, \cdots, x_{n-2}, w, v.$$

If x_i is any member of $N(v)$ then x_{i-1} cannot be a member of $N(w)$, because

$$v, x_1, x_2, \cdots, x_{i-1}, w, x_{n-2}, x_{n-3}, \cdots, x_i, v$$

would be a Hamilton cycle in G. So each of the $d(v)$ vertices adjacent to v in G must be preceded in the cycle by vertices not adjacent to w, and none of these vertices can be w itself. So there are at least $d(v) + 1$ vertices in G which are not adjacent to w. So there are at least $d(w) + d(v) + 1$ vertices in G, whence

$$d(v) + d(w) \leq n - 1,$$

a contradiction.   □

Corollary 2.3.1 *If G is a graph with n vertices, $n \geq 3$, and every vertex has degree at least $\frac{n}{2}$, then G is Hamiltonian.*   □

Theorem 2.3 was first proven by Ore [122] and Corollary 2.3.1 some years earlier by Dirac [60]. Both can in fact be generalized into the following result of Pósa [129]: a graph with n vertices, $n \geq 3$, has a Hamiltonian cycle provided the number of vertices of degree less than or equal to k does not exceed k, for each k satisfying $1 \leq k \leq \frac{n-1}{2}$.

Another famous property, which sounds rather like Hamiltonicity but is in fact much easier to study, was introduced by Euler [63] in what is recognized as the first paper on graph theory. In his honor we define an *Euler walk* in a graph to be a closed walk in which every edge occurs exactly once. Graphs which possess Euler walks are called *Eulerian*.

Theorem 2.4 *A graph G has an Euler walk if and only if it is connected and every vertex has even degree.*

Proof. Clearly no disconnected graph can contain an Euler walk. Given an Euler walk

$$x_0, a_1, x_1, a_2, x_2, \cdots, a_n, x_n,$$

where $x_n = x_0$, suppose the occurrences of x_i are $x_{i_1} = x_{i_2} = \cdots = x_{i_k}$. Then the edges touching x_i are $a_{i_1}, a_{i_1+1}, a_{i_2}, a_{i_2+1}, \cdots, a_{i_k}, a_{i_k+1}$. This means $d(x_i) = 2k$, which is even. (If x_{n+1} arises, interpret it as x_1.) So every degree is even. Therefore the conditions are necessary.

Now assume G is connected and every vertex is of even degree. We present an algorithm which produces an Euler walk. First select a vertex x_0, and define $G_0 = G$. Now proceed: if x_i has degree greater than 0 in G_i, select an edge a_{i+1} in G_i which touches x_i; define x_{i+1} to be the other endpoint of a_{i+1}, and

$G_{i+1} = G_i - a_{i+1}$. This process can only stop when x_i has degree 0 in G_i, and that can occur only when $x_i = x_0$: Clearly x_i has odd degree in G_i otherwise. So the algorithm results in a closed walk. Once the algorithm stops, leaving G_n say, select any vertex in $\{x_0, x_1, \cdots, x_n\}$ whose degree in G_n is non-zero, and repeat the algorithm; if x_i is chosen, then construct the walk

$$x_0, a, x_1, a_2, x_2, \cdots, x_i, \text{ (2nd walk) }, x_i, a_i, \cdots, x_n.$$

Continue. The result is an Euler walk. □

The corresponding result concerning open walks is left as an exercise (see Exercise 2.6).

Exercises 2

2.1 Prove that a spanning subgraph of $K_{m,n}$ can have a Hamilton cycle only if $m = n$.

2.2 Prove that for every n there exists a graph on n vertices such that, for any two nonadjacent vertices x and y, $d(x) + d(y) \geq n - 1$.

2.3 G is a graph with n vertices; x and y are nonadjacent vertices of G satisfying $d(x) + d(y) \geq n$. Prove that $G + uv$ is Hamiltonian if and only if G is Hamiltonian. [23]

2.4 G is a graph with n vertices; $n \geq 3$. Prove that if G has at least $\frac{n^2 - 3n + 6}{2}$ edges then G is Hamiltonian.

2.5 Prove that a graph is a tree if and only if each of its vertices is a cutpoint.

2.6 An open Euler walk in a graph G is a walk $x_0, a_1, x_1, a_2, \cdots, x_n$ which contains every edge precisely once, in which $x_0 \neq x_n$. Prove that G contains an open Euler walk if and only if G is connected and contains precisely two vertices of odd degree.

3

ONE-FACTORS AND
ONE-FACTORIZATIONS

If G is any graph, then a *factor* or *spanning subgraph* of G is a subgraph with vertex-set $V(G)$. A *factorization* of G is a set of factors of G which are pairwise *edge-disjoint* — no two have a common edge — and whose union is all of G.

Every graph has a factorization, quite trivially: since G is a factor of itself, $\{G\}$ is a factorization of G. However, it is more interesting to consider factorizations in which the factors satisfy certain conditions. In particular a *one-factor* is a factor which is a regular graph of degree 1. In other words, a one- factor is a set of pairwise disjoint edges of G which between them contain every vertex. A *one-factorization* of G is a decomposition of the edge-set of G into edge-disjoint one-factors.

Another approach to the study of one-factors is through matchings. A *matching* between sets X and Y is a set of ordered pairs, one member from each of the two sets, such that no element is repeated. Such a matching is a set of disjoint edges of the $K_{m,n}$ with vertex-sets X and Y. The matching is called *perfect* if every member occurs exactly once, so a perfect matching is a one-factor in a complete bipartite graph. One can then define a matching in any graph to be a set of disjoint edges in that graph; in this terminology "perfect matching" is just another word for "one-factor".

It is sometimes useful to impose an ordering on the set of one-factors in a one-factorization, or a direction on the edges of the underlying graph. (See for example the applications in Chapter 5.) In those cases the one-factorization will be called *ordered* or *oriented* respectively.

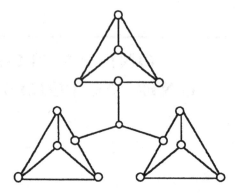

Figure 3.1   The smallest cubic graph without a one-factor.

Not every graph has a one-factor. In Chapter 6 we shall give a necessary and sufficient condition for the existence of a one-factor in a general graph. For the moment we note the obvious necessary condition that a graph with a one-factor must have an even number of vertices. However, this is not sufficient; Figure 3.1 shows a 16-vertex graph without a one-factor (see Exercise 3.1).

In order to have a one-factorization, a graph not only needs an even number of vertices, but it must also be regular: if G decomposes into d disjoint one-factors, then every vertex of G must lie on precisely d edges. The following theorem shows that these conditions are not sufficient.

Theorem 3.1 *A regular graph with a bridge cannot have a one-factorization (except for the trivial case where the graph is itself a one-factor).*

Proof. Consider a regular graph G of degree d, $d > 1$, with a bridge $e = uv$; in $G - e$, label the component which contains u as E and label the component which contains v as F. The fact that e is a bridge implies that E and F are distinct. Suppose G is the edge-disjoint union of d one-factors, $G_1, G_2, \cdots, G_d$; and say G_1 is the factor which contains e. Now e is the only edge which joins a vertex of E to a vertex of F, so every edge of G_2 with one endpoint in E has its other endpoint in E. So G_2 contains an even number of vertices of E. Since G_2 contains every vertex of the original graph, it contains every vertex of E; so E must have an even number of vertices.

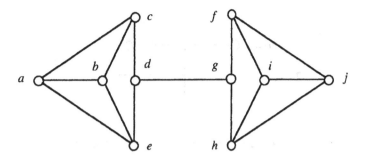

Figure 3.2 M, the smallest cubic graph without a one-factorization.

On the other hand, consider $G_1 \backslash E$. This contains a number of edges and the isolated vertex u, since e is in G_1. So $G_1 \backslash E$ has an odd number of vertices, and accordingly E has an odd number of vertices. We have a contradiction. $\square$

This theorem can be used, for example, to show that the graph M of Figure 3.2 has no one-factorization, although it is regular and possesses the one-factor $\{ac, be, dg, fi, hj\}$. However, it clearly does not tell all the story: the Petersen graph P, shown in Figure 3.3, has no one-factorization, but it also contains no bridge. The Petersen graph *does* contain a one-factor, however. In fact Petersen [125] showed that every bridgeless cubic graph contains a one-factor. We shall give a proof in Chapter 6.

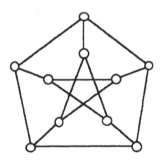

Figure 3.3 P, the Petersen graph.

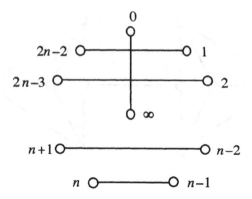

Figure 3.4 The factor F_0 in $\mathcal{G}\mathcal{K}_{2n}$.

If the degree increases with the number of vertices, the situation is different. It has been conjectured that a regular graph with $2n$ vertices and degree greater than n will always have a one-factorization; this has only been proven in a very few cases, such as degree $2n - 4$, degree $2n - 5$, and degree at least $12n/7$ (for further details see [132], [37] and Chapter 19). On the other hand, Theorem 3.1 allows us to show that there are regular graphs with degree near to half the number of vertices, which do not have one-factorizations — see Exercise 3.7.

However, we can prove the existence of one-factorizations in many classes of graphs. Of basic importance are the complete graphs. There are many one-factorizations of K_{2n}. We present one which is usually called $\mathcal{G}\mathcal{K}_{2n}$. To understand the construction, look at Figure 3.4. This represents a factor which we shall call F_0. To construct the factor F_1, rotate the diagram through a $(2n-1)$-th part of a full revolution. Similar rotations provide $F_2, F_3, \cdots, F_{2n-2}$. We shall present an algebraic version of this construction now, but we revert to the geometric interpretation in Chapter 10, when we generalize the construction of $\mathcal{G}\mathcal{K}_{2n}$ to the case of construction from a starter.

Theorem 3.2 *The complete graph K_{2n} has a one-factorization for all n.*

Proof. We label the vertices of K_{2n} as $x_\infty, x_0, x_1, x_2, \cdots, x_{2n-2}$. The spanning subgraph F_i is defined to consist of the edges

$$x_\infty x_i, x_{i+1} x_{i-1}, \cdots, x_{i+j} x_{i-j}, \cdots, x_{i+n-1} x_{i-n+1} \tag{3.1}$$

where the subscripts other than ∞ are treated as integers modulo $2n - 1$. Then F_i is a one-factor: every vertex appears in the list (3.1) exactly once. We prove that $\{F_0, F_1, \cdots, F_{2n-2}\}$ form a one-factorization. First, observe that every edge involving x_∞ arises precisely once: $x_\infty x_i$ is in F_i. If neither p nor q is ∞, then we can write $p + q = 2i$ in the arithmetic modulo $2n - 1$, because either $p + q$ is even or $p + q + 2n - 1$ is even. Then $q = i - (p - i)$, and $x_p x_q$ is $x_{i+j} x_{i-j}$ in the case $j = p - i$. Since i is uniquely determined by p and q, this means that $x_p x_q$ belongs to precisely one of the F_i. So $\{F_0, F_1, \cdots, F_{2n-2}\}$ is the required one-factorization. □

Further general examples are given in Exercises 3.2 and 3.4. Another important case is the family of cycles C_n: these have a one-factorization if and only if n is even. This fact will be useful — for example, one common way to find a one-factorization of a cubic graph is to find a spanning subgraph which is a union of disjoint even cycles: a Hamilton cycle will suffice. The complement of this subgraph is a one-factor, so the graph has a one-factorization. Reversing this reasoning, the union of two disjoint one-factors is always a union of disjoint even cycles; if the one-factors are not disjoint, the union consists of some even cycles and some isolated edges (the common edges of the two factors).

The use of unions in the preceding paragraph can be generalized. If G and H both have one-factorizations, then so does $G \oplus H$, the factorization being formed by listing all factors in the factorizations of each of the component graphs. At the other extreme, if G has a one-factorization, then so does nG. However, care must be exercised. If G and H have some common edges, nothing can be deduced about the factorization of $G \cup H$ from factorizations of G and H.

The complete bipartite graph $K_{n,n}$ is easily shown to have a one-factorization. If $K_{n,n}$ is defined to have vertex-set $\{1, 2, \cdots, 2n\}$ and edge-set $\{(x, y) : 1 \leq x \leq n, n + 1 \leq y \leq 2n\}$, then the factors $F_1, F_2, \cdots, F_n$, defined by

$$F_i = \{(x, x + n + i) : 1 \leq x \leq n\}$$

(where $x + n + i$ is reduced modulo n to lie between $n + 1$ and $2n$), form a one-factorization. This will be called the *standard factorization* of $K_{n,n}$.

One-factorizations of complete bipartite graphs are equivalent to the arrays called *Latin squares*. A Latin square of side n is an $n \times n$ array with entries from $\{1, 2, \cdots, n\}$, in which every row and every column contains each symbol precisely once. Given a Latin square $L = (\ell_{ij})$ of side n, one can construct a one-factorization of $K_{n,n}$ as follows: the j-th factor consists of all the pairs $(i, \ell_{ij} + n)$. (One could also exchange the roles of rows and columns, or columns and array entries, and so on.) Conversely, the existence of a one-factorization of $K_{n,n}$ implies the existence of a Latin square of side n.

One could view K_{2n} as the union of three graphs: two disjoint copies of K_n and a copy of $K_{n,n}$. When n is even, this gives rise to a one-factorization in an obvious way: $n-1$ of the factors consist of the union of one factor from each of the K_n, and the other n factors are a one-factorization of the $K_{n,n}$. Such factorizations are called *twin* factorizations, and arise in Chapter 12. A slight generalization covers the case of odd n. We present here a particular case which is one of the standard one-factorizations.

First suppose n is even. Label the vertices of K_{2n} as

$$1_1, 2_1, \cdots, n_1, 1_2, 2_2, \cdots, n_2.$$

Write $P_{\alpha,1}, P_{\alpha,2}, \cdots, P_{\alpha,n-1}$ for the factors in $\mathcal{G}\mathcal{K}_n$ with the symbols $x_1, x_2,$ $\cdots, x_{n-2}, x_0, x_\infty$ replaced by $1_\alpha, 2_\alpha, \cdots, n_\alpha$ consistently, and write $H_n, H_{n+1},$ $\cdots, H_{2n-1}$ for the factors in the standard factorization of $K_{n,n}$ with $1, 2, \cdots, 2n$ replaced by $1_1, 2_1, \cdots, n_1, 1_2, \cdots, n_2$. If we define $H_i = P_{1,i} \cup P_{2,i}$ for $1 \leq i \leq n-1$, then $H_1, H_2, \cdots, H_{2n-1}$ make up the factorization $\mathcal{G}\mathcal{A}_{2n}$.

Now suppose n is odd. We use the same vertex-set. Let $P_{\alpha,1}, P_{\alpha,2}, \cdots, P_{\alpha,n}$ be the factors of $\mathcal{G}\mathcal{K}_{2n}$, where this time the vertices are $1_\alpha, 2_\alpha, \cdots, n_\alpha, \infty_\alpha$. We define

$$J_i = [P_{1,i} \cup P_{2,i} \cup \{(i_1, i_2)\}] \backslash \{(\infty_1, i_1), (\infty_2, i_2)\}.$$

Then $J_1, J_2, \cdots, J_n$ are one-factors of K_{2n}, and they contain all the edges of the two copies of K_n together with the edges of the factor

$$(1_1, 1_2), (2_1, 2_2), \cdots, (n_1, n_2)$$

of $K_{n,n}$. But this is one of the factors of the standard factorization. We write $J_{n+1}, J_{n+2}, \cdots, J_{2n-1}$ for the remaining factors in $K_{n,n}$, written in terms of our vertex-set. Then $\{J_1, J_2, \cdots, J_{2n-1}\}$ is $\mathcal{G}\mathcal{A}_{2n}$. (For consistency, we shall always take J_{n+k} to consist of all the edges $(i_1, (i+k)_2)$ in discussions of $\mathcal{G}\mathcal{A}_{2n}$, where $i+k$ is reduced modulo n if necessary.)

Theorem 3.3 $\mathcal{G}\mathcal{A}_{2n}$ *is a one-factorization of* K_{2n} *for all* n.   □

Another important result concerns the graph $G(n, w)$, which is defined to have the integers modulo $2w$ as its vertices and edges xy whenever $w - n < x - y < w + n$. This graph has been studied by several authors ([84], [141], [140]). We shall follow the discussion in [140]. We write Q_d for the subgraph of $G(n,w)$ formed by the edges xy where $x - y \equiv d \pmod{2w}$. Clearly $G(n,w)$ equals the disjoint union

$$G(n,w) = Q_{w-n+1} \cup Q_{w-n+2} \cup \cdots \cup Q_w.$$

Q_w is a single one-factor. If we write δ for the greatest common divisor $\gcd(d, 2w)$, then Q_d consists of the δ cycles

$$i, d + i, 2d + i, \cdots, i - d \pmod{2w}$$

for $i = 0, 1, \cdots, \delta - 1$. When $\frac{2w}{\delta}$ is even, these cycles are even, so Q_d has a one-factorization. In particular, Q_d has a one-factorization whenever d is odd, and Q_{w-1} has a one-factorization. Various possibilities exist according to the parities of w and d, but in every case $G(n,w)$ can be decomposed into some graphs which we know to have one-factorizations together with the $Q_{2x} \cup Q_{2x+1}$ where $\delta = \gcd(2x, 2w)$ is such that $\frac{2w}{\delta}$ is odd.

We now show (following [140]) that $Q_{2x} \cup Q_{2x+1}$ splits into four one-factors. We start with a decomposition $F_1 \cup F_2$ of Q_{2x+1} into two one-factors:

$$
\begin{aligned}
F_1 &= \big\{ (i, 2x + i + 1), (4x + i + 2, 6x + i + 3) \\
&\qquad \cdots (i - 4x - 2, i - 2x - 1) : i \text{ odd} \big\} \\
&\cup \big\{ (i - 2x - 1, i), (2x + i + 1, 4x + i + 2) \\
&\qquad \cdots (i - 6x - 4, i - 4x - 2) : i \text{ even} \big\}
\end{aligned}
$$

while F_2 is obtained by exchanging the conditions "i odd" and "i even" in the above description. We use two further factors, F_3 and F_4, defined as

$$
\begin{aligned}
F_3 &= \big\{ (2x + i, 4x + i), (6x + i, 8x + i) \cdots (-4x + i, -2x + i) \\
&\quad (4x + i + 1, 6x + i + 1), (8x + i + 1, 10x + i + 1) \\
&\quad \cdots (-2x + i + 1, i + 1), (i, 2x + i + 1) : i \text{ odd} \big\}
\end{aligned}
$$

$$
\begin{aligned}
F_4 &= \big\{ (i, 2x + i), (4x + i, 6x + i) \cdots (-6x + i, -4x + i) \\
&\quad (2x + i + 1, 4x + i + 1), (6x + i + 1, 8x + i + 1) \\
&\quad \cdots (-4x + i + 1, -2x + i + 1), (i - x, i + 1) i \text{ odd} \big\}
\end{aligned}
$$

These four factors do not form a one-factorization: the edges in the set A,

$$A = \big\{ (i, 2x + 1 + i), (i - 2x, i + 1) : i \text{ odd} \big\},$$

occur in F_1 and also in $F_3 \cup F_4$, while the edges in B,

$$B = \big\{ (i, i - 2x), (i + 1, 2x + i + 1) : i \text{ odd} \big\},$$

are omitted. But $(F_1 \backslash A) \cup B$ is a one-factor, and it, together with F_2, F_3 and F_4, forms the required one-factorization. We have proven:

Theorem 3.4 $G(n, w)$ *has a one-factorization for all n and w.*   □

Although our main interest is in one-factorizations, we will use some other factorizations of complete graphs, which we present in the next two Theorems.

Theorem 3.5 *If v is odd, then K_v can be factored into $\frac{v-1}{2}$ Hamilton cycles. If v is even, then K_v can be factored into $\frac{v}{2} - 1$ Hamilton cycles and a one-factor.*

Proof. First, suppose v is odd: say $v = 2n+1$. If K_v has vertices $0, 1, 2, \cdots, 2n$, then a suitable factorization is $Z_1, Z_2, \cdots, Z_n$, where

$$Z_i = (0, i, i+1, i-1, i+2, i-2, \cdots, i+j, i-j, \cdots, i+n, 0).$$

(If necessary, reduce integers modulo $2n$ to the range $\{1, 2, \cdots, 2n\}$.)

In the case where v is even, let us write $v = 2n+2$. We construct an example for the K_v with vertices $\infty, 0, 1, 2, \cdots, 2n$. The factors are the cycles $Z_1, Z_2, \cdots, Z_n$, where

$$Z_i = (\infty, i, i-1, i+1, i-2, i+2, \cdots, i+n-1, \infty),$$

and the one-factor

$$(\infty, 0), (1, 2n), (2, 2n-1), \cdots, (n, n+1)$$ □

The nearest thing to a one-factor in K_{2n-1} is a set of $n-1$ edges which cover all but one vertex. Such a structure is called a *near-one-factor*. A set of near-one-factors which covers every edge precisely once is called a *near-one-factorization*.

Theorem 3.6 K_{2n-1} *has a near-one-factorization for every n.*

Proof. Consider a one-factorization of the K_{2n} with vertex-set $\{1, 2, \cdots, 2n\}$. Delete from it all the edges $(1, 2n), (2, 2n), \cdots, (2n-1, 2n)$. The result is a near-one-factorization of the K_{2n-1} with vertex-set $\{1, 2, \cdots, 2n-1\}$.   □

It is in fact obvious that any near-one-factorization of K_{2n-1} can be converted to a one-factorization of K_{2n} by adding a vertex ∞ and joining ∞ to the isolated vertex in each factor. It follows that each vertex appears precisely once as an isolate in a near-one-factorization. (This also follows immediately from adding degrees.)

Theoretical considerations aside, it is often important to be able to construct a one-factorization (or a set of one-factorizations) of K_{2n}, for a particular value $2n$. A hill-climbing algorithm was described by Dinitz and Stinson [55]. In order to use the hill-climbing approach, it is necessary to formulate the search for a one-factorization as an optimization problem. One first defines a *pairing* to be a set $\mathcal{F}$ of pairs of the form $\{f_i, (x, y)\}$, where each edge (x, y) of K_{2n} occurs as the latter entry in at most one of the pairs, and where $\mathcal{F}$ contains no two pairs $\{f_i, (x, y)\}$ and $\{f_i, (x, z)\}$ where $y \neq z$. The *cost* $c(\mathcal{F})$ of a pairing $\mathcal{F}$ is $n(2n - 1) - |\mathcal{F}|$. Then $\mathcal{F}$ is a one-factorization if and only if $c(\mathcal{F}) = 0$.

The algorithm proceeds as follows. Assume a pairing $\mathcal{F}$ has been given. First, choose a vertex x which belongs to fewer than $2n - 1$ of the elements $\{f_i, (x, y)\}$ of $\mathcal{F}$. Then choose any f_i such that $\{f_i, (x, y)\}$ is not in $\mathcal{F}$ for any y. Next choose a y such that $\{f_j, (x, y)\}$ is not in $\mathcal{F}$ for any i. Finally,

(i) if there is no pair $\{f_i, (z, y)\}$ in $\mathcal{F}$ for any z, then add $\{f_i, (x, y)\}$ to $\mathcal{F}$;

(ii) otherwise, if $\{f_i, (z, y)\}$ belongs to $\mathcal{F}$, $z \neq x$, then delete $\{f_i, (z, y)\}$ from $\mathcal{F}$ and add $\{f_i, (x, y)\}$.

This always gives a new pairing whose cost is no greater than that of $\mathcal{F}$. If no improvement occurs, one can try another heuristic: choose any f_i and choose two vertices x and y, each of which occurs fewer than $2n - 1$ times in $\mathcal{F}$. Select an f_i such that no pair $\{f_i, (x, z)\}$ or $\{f_i, (y, t)\}$ ever occurs in $\mathcal{F}$. Then:

(i) if $\{f_j, (x, y)\}$ does not occur in $\mathcal{F}$ for any j, then add $\{f_i, (x, y)\}$ to $\mathcal{F}$;

(ii) if $\{f_j, (x, y)\}$ occurs in $\mathcal{F}$, replace $\{f_j, (x, y)\}$ by $\{f_i, (x, y)\}$.

There is no guarantee that repeated applications of these heuristics will produce a one-factorization; but Dinitz and Stinson [55] report no failures in over a million trials.

Exercises 3

3.1 Prove that the graph N of Figure 3.1 contains no one-factor.

3.2 The n-cube Q_n is defined as follows. Q_1 consists of two vertices and one edge; Q_2 is the cycle C_4; in general Q_n is formed by taking two copies of Q_{n-1} and joining each vertex in one copy to the corresponding vertex in the other copy.

(i) How many vertices does Q_n have?

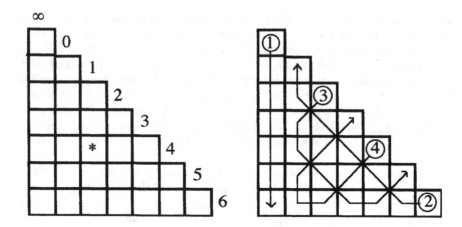

Figure 3.5   Staircase factorization.

(ii) Prove that Q_n is regular. What is the degree?

(iii) Prove that Q_n has a one-factorization.

(iv) Prove that Q_n has a Hamilton cycle, for $n > 1$.

3.3 The following construction is due to F. Bileski. A *staircase diagram* of order $2n$ consists of $2n - 1$ rows of cells, where row i contains $i + 1$ cells. Each rows is labeled by its number, placed to the right of the rightmost cell, and this label also labels the column below it. This is illustrated for $2n = 8$ on the left of Figure 3.5; the asterisk is in cell $(4,1)$. The first column is column ∞. Paths are inserted as follows. Path 1 is vertical. Other odd-numbered paths $(3,5,\cdots)$ are constructed by the rules:

Southwest as far as possible
South one step
Southeast as far as possible
East one step
Northeast as far as possible

(where "as far as possible" means "go until the diagram will take you outside the diagram or into path 1". Path 3 starts on the third step (position $(1,2)$), path 5 on the fifth, and so on. The even-numbered paths have the same description, except that the sequence of directions is

Southwest, West, Northwest, North, Northeast.

In other words, always make a 45° turn in the appropriate direction. Path 2 starts on step $2n - 1$ (the last step, position $(2n - 3, 2n - 2)$), path 4 on step $2n-3$, et cetera. The paths are shown on the right in Figure 3.5. Prove

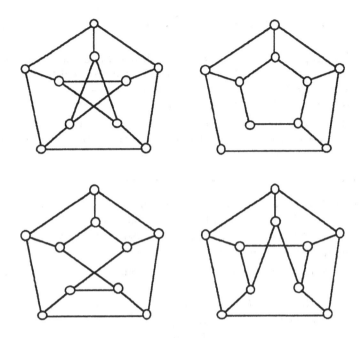

Figure 3.6   Four cubic graphs on ten points.

that the cells on each path form a one-factor of the K_{2n} with vertex-set
$\{\infty, 0, 1, \cdots, 2n_2\}$ and that the set of $2n-1$ factors is a one-factorization.
[165]

3.4 Which of the graphs in Figure 3.6 have one-factorizations?

3.5 What is the number of distinct one-factors in K_{2n}?

3.6 A *symmetric idempotent* Latin square A is a Latin square $A = (a_{ij})$ in
which $a_{ii} = i$ for each i and $a_{ij} = a_{ji}$ for each i and j. Prove that
the existence of a symmetric idempotent Latin square of side $2n - 1$ is
equivalent to the existence of a one-factorization of K_{2n}. (This will be
discussed in the next Chapter.)

3.7 Consider a one-factorization of K_{2t}. A graph H is formed by taking the
union of d of the factors in the one-factorization. If d is odd, G is formed
from H by deleting $(d-1)/2$ of the edges from one of the factors which
constituted H, introducing a new vertex z, and — for every edge xy which
was deleted from H — introducing edges xz and yz.

    (i) Prove that G has $2t$ vertices of degree d and one vertex of degree $d-1$.

    (ii) If a graph has $2t+1$ vertices, $2t$ of degree d and one of degree $d-1$, we shall call it a $G(d, 2t+1)$. Prove that if there exist a $G(d, 2s+1)$ and a $G(d, 2t+1)$ then there is a graph on $2s+2t+2$ vertices which is regular of degree d and has a bridge.

    (iii) Hence show that there is a connected regular graph of degree d on $2n$ vertices which has no one-factor whenever $d \le n-2$ when n is odd and whenever $d \le n-3$ when n is even.

3.8 Find a one-factorization of K_8 which is different from $\mathcal{G}K_8$.

3.9 Suppose $\mathcal{F}$ and $\mathcal{G}$ are near-one-factorizations of the K_{2n+1} with vertex-set $\{0, 1, \cdots 2n\}$. Write F_i and G_i for the near-factors in $\mathcal{F}$ and $\mathcal{G}$ respectively which omit vertex i. Define $\mathcal{F}$ and $\mathcal{G}$ to be *disjoint* if it never happens that F_i and G_i have a common edge, for any i.

    (i) Prove that there do not exist two disjoint near-one-factorizations of K_5.

    (ii) Prove that there can never be more than $2n-1$ mutually disjoint near-one-factorizations of K_{2n+1}.

    (iii) A set of $2n-1$ mutually disjoint near-one-factorizations of K_{2n+1} is called *large*. Find a large set of mutually disjoint near-one-factorizations of K_7.

3.10 Consider the subgraphs Q_d defined in the discussion of $G(n, w)$. Verify that Q_d has a one-factorization whenever $\gcd(d, 2w)$ is odd, and therefore that Q_d has a one-factorization whenever d is odd.

4

ORTHOGONAL
ONE-FACTORIZATIONS

There are a number of applications of one-factorizations in the theory of combinatorial designs. In general this topic is too big to discuss here, but we shall explore a couple of examples. In this chapter we look at the applications concerning Latin squares; one-factorizations and block designs are discussed in Chapter 9.

The correspondence between Latin squares and one-factorizations of complete bipartite graphs was mentioned in Chapter 3. Another relationship exists between factorizations and a special class of Latin squares. A Latin square $L = (\ell_{ij})$ is called *symmetric* if $\ell_{ij} = \ell_{ji}$ for every i and j. It is easy to see that a symmetric Latin square of odd side r must have each of the symbols $1, 2, \cdots, r$ exactly once on its main diagonal. (See Exercise 4.1.)

Suppose L is a symmetric Latin square of side $r = 2n - 1$. Write S_k for the set of all the pairs (i, j) such that $\ell_{ij} = k$. From the symmetry of L it follows that S_k contains entry (i, j) if and only if it contains (j, i), so we can interpret the elements of S_k as edges of a graph. In fact, it is clear from the Latin square properties of L that S_k is a near-one-factor of the K_{2n-1} with vertices $\{1, 2, \cdots, 2n - 1\}$, and that $\{S_1, S_2, \cdots, S_{2n-1}\}$ form a near-one-factorization. The vertex i is omitted from S_k, where k lies in the i-th diagonal entry of L. Alternatively one could use L to define a one-factorization of the K_{2n} based on $\{0, 1, \cdots, 2n - 1\}$: simply add the pair $(0, i)$ to S_k, where i is the vertex missing from S_k.

The above construction is easily reversed, so that each one-factorization of K_{2n} or near-one-factorization of K_{2n-1} gives rise to a symmetric Latin square of side $2n - 1$. There will, in fact, be $(2n - 1)!$ such Latin squares for each factorization, because the $2n - 1$ factors could be taken in any order; it is usual to permute the symbols in a symmetric Latin square of odd order so as to achieve diagonal $(1, 2, \cdots, 2n - 1)$, and if this is done then symmetric Latin squares of order $2n - 1$ are equivalent to one-factorizations of K_{2n}. When the diagonal is ordered in this way, the Latin square is called *idempotent*.

One very important property of Latin squares is *orthogonality*. Latin squares L and M of side n are called *orthogonal* if they have the following property: in the n positions where L has the symbol x, M has n different symbols, for every possible value of x. To put it another way, writing S for $\{1, 2, \cdots, n\}$,

$$\{m_{ij} : \ell_{ij} = x\} = S$$

for all x in S. This is equivalent to saying

$$\{(\ell_{ij}, m_{ij}) : 1 \leq i \leq n, 1 \leq j \leq n\} = S \times S$$

and, by a trivial counting argument, the set $\{(\ell_{ij}, m_{ij})\}$ contains no repetitions: every possible ordered pair occurs exactly once. This interpretation makes it clear that orthogonality is symmetrical: L is orthogonal to M if and only if M is orthogonal to L.

Given Latin squares L and M, write

$$\begin{aligned} \mathcal{L} &= \{L_1, L_2, \cdots, L_n\}, \\ \mathcal{M} &= \{M_1, M_2, \cdots, M_n\} \end{aligned}$$

for the corresponding one-factorizations of $K_{n,n}$:

$$L_j = \{(i, \ell_{ij} + n)\}, \quad M_k = \{(i, m_{ik} + n)\}.$$

Then it is easy to see that L and M are orthogonal if and only if $L_j \backslash M_k$ has precisely one element. We will say that $\mathcal{L}$ and $\mathcal{M}$ *are orthogonal* in this case. To generalize this, we will say that two one-factors of a graph are orthogonal if they have at most one edge in common; two one-factorizations are orthogonal if each factor of one is orthogonal to each factor of the other.

Suppose $\mathcal{L} = \{L_1, L_2, \cdots, L_r\}$ and $\mathcal{M} = \{M_1, M_2, \cdots, M_r\}$ are two orthogonal one-factorizations of a graph G. Then one can construct an $r \times r$ array whose entries are either empty or contain unordered pairs of vertices of G, as follows. The (i, j) entry contains $L_i \cap M_j$. The definition of orthogonality implies that no cell contains more than one unordered pair. If the graph G is $K_{n,n}$, then the array produced corresponds to a pair of orthogonal Latin squares in an obvious way: if the (i,j) entry is $\{\ell_{ij}, s_{ij}\}$ where $\ell_{ij} \leq s_{ij}$ then $1 \leq \ell_{ij} \leq n$ and $n + 1 \leq s_{ij} \leq 2n$, and the arrays L and M, where $L = (\ell_{ij})$ and $M = (m_{ij})$, $m_{ij} = s_{ij} - n$, are orthogonal Latin squares. In the case where G is the complete graph K_{2n}, the array is called a *Room square*. A Room square R based on a $2n$-set S is a square array of side $2n - 1$ whose cells are empty or contain unordered pairs of elements of S, such that every row and every column contains every element of S precisely once, and such that every unordered pair of elements of S occurs precisely once in the array.

Theorem 4.1 [27] *There exists a pair of orthogonal Latin squares of every side greater than 2, except side 6.* □

Theorem 4.2 [155] *There exists a Room square of every odd side except 3 and 5.* □

In both cases the above Theorems are exact — orthogonal Latin squares of side 2 or 6, and Room squares of side 3 or 5, are impossible. In the Latin square case, side 2 is trivial; side 6 was eliminated by Tarry using an exhaustive computation [147] and theoretically by Stinson [144]. In a Room square of side 3, the entry 01 would appear somewhere, say in the $(1, 1)$ position; to complete row 1, 23 must lie in position $(1, 2)$ or $(1, 3)$, but to complete column 1 the same entry must lie in position $(2, 1)$ or $(3, 1)$, and this contradicts the fact that each pair can appear only once. Room squares of side 5 are eliminated by a similar, slightly longer, argument; this is left as an exercise.

It is often useful to have a set of Latin squares which are *pairwise orthogonal* — each one orthogonal to each other one. It is easy to see that no set of pairwise orthogonal Latin squares of side n can have more than $n - 1$ members (relabel the entries so that each square has first row $12 \cdots n$; the $(2, 1)$ entries must all be different and cannot equal 1). The bound $n - 1$ can be attained when n is a prime power; it is widely conjectured that no other cases exist, and this is one of the most interesting questions of combinatorial design theory. The situation is:

Theorem 4.3 *There exists a set of $n - 1$ pairwise orthogonal Latin squares of side n if and only if there exists a finite projective plane of parameter n. Such a plane is known whenever n is a prime power.* □

$\mathcal{F}_1$	01	23	45	67	89
	02	13	46	58	79
	03	12	47	39	68
	04	16	25	39	78
	05	18	24	37	69
	06	19	27	35	48
	07	15	28	36	49
	08	17	29	34	56
	09	14	26	38	57

$\mathcal{F}_2$	01	29	36	48	57
	02	15	34	69	78
	03	16	28	45	79
	04	17	26	35	89
	05	14	27	39	68
	06	12	37	49	58
	07	19	25	38	46
	08	13	24	59	67
	09	18	23	47	56

$\mathcal{F}_3$	01	26	39	47	58
	02	14	37	56	99
	03	17	25	48	69
	04	18	27	36	59
	05	19	28	34	67
	06	15	24	38	79
	07	13	29	45	68
	08	16	23	49	67
	09	12	25	46	78

$\mathcal{F}_4$	01	25	34	68	79
	02	18	35	99	67
	03	15	27	46	89
	04	13	28	57	69
	05	16	29	38	47
	06	14	23	59	78
	07	12	39	48	56
	08	19	26	37	45
	09	17	24	36	58

Table 4.1 Four orthogonal one-factorizations of K_{10}.

Even when the bound is not attained, quite a number of results on pairwise orthogonal Latin squares are known. For example, the work of several authors has terminated in

Theorem 4.4 *There is a set of 3 pairwise orthogonal Latin squares of order n for all $n > 10$.*   □

For further details and more extensive references see, for example, [162].

Pairwise orthogonal one-factorizations of complete graphs have also been studied, although there are fewer results known than in the complete bipartite case. If symmetric Latin squares correspond to orthogonal one-factorizations, then they are called *orthogonal symmetric Latin squares*. One can speak of a set of k *pairwise orthogonal symmetric Latin squares* of side $2n - 1$ (or k-POSLS($2n - 1$)). All members of such a set are of course idempotent.

r	$\nu(r) \geq$		r	$\nu(r) \geq$		r	$\nu(r) \geq$		r	$\nu(r) \geq$
1	$= 1$		27	13		53	17		79	39
3	$= 1$		29	13		55	5		81	5
5	$= 1$		31	15		57	5		83	41
7	$= 3$		33	5		59	29		85	5
9	$= 4$		35	5		61	21		87	5
11	5		37	15		63	5		89	11
13	5		39	5		65	5		91	5
15	4		41	9		67	33		93	5
17	5		43	21		69	5		95	5
19	9		45	5		71	35		97	5
21	5		47	25		73	9		95	5
23	11		49	5		75	5		101	31
25	7		51	5		77	5			

Table 4.2 Known numbers of pairwise orthogonal symmetric Latin squares.

Theorem 4.5 *There can exist no k-POSLS$(2n - 1)$ when $k > 2n - 3$, except in the trivial case $n = 1$.*

Proof. Suppose $M_1, M_2, \cdots, M_k$ are pairwise orthogonal symmetric Latin squares of side $2n - 1$. Write m_i for the (1,2) entry of M_i. Then m_i cannot equal 1 or 2 for any i, by the idempotence of M_i and the Latin property, and the m_i are all different members of $\{3, 4, \cdots, 2n - 1\}$. So there can be at most $2n - 3$ of them. $\square$

No case is known of $2n - 3$ pairwise orthogonal symmetric Latin squares of side $2n - 1$ except in the trivial case $2n - 1 = 3$; in fact no case is known where the number of squares is greater than $n - 1$. Let us write $\nu(r)$ for the largest value ν such that ν-POSLS(r) can exist. We know that $\nu(1) = \nu(3) = \nu(5) = 1$, $\nu(7) = 3$ and $\nu(9) = 4$, by complete searches. No other exact value is known. Theorem 4.2 implies that $\nu(r) \geq 2$ for all odd r greater than 5.

A set of four pairwise orthogonal one-factorizations of K_{10} was exhibited in [58], and is shown in Table 4.1. This set has an interesting structure. It is unique up to isomorphism. In the notation of the Table, $\mathcal{F}_2$, $\mathcal{F}_3$ and $\mathcal{F}_4$ are isomorphic to each other, but not to $\mathcal{F}_1$. The set of factorizations has an automorphism group of order 3, generated by the permutation $(013)(476)(598)$, which leaves $\mathcal{F}_1$ invariant and maps $\mathcal{F}_2 \mapsto \mathcal{F}_3 \mapsto \mathcal{F}_4$.

Various results are known which improve the lower bound on the value $\nu(r)$ (see, for example, Corollary 10.4.1, and Corollary 10.6.1 and the discussion following it). Table 4.2 shows the lower bounds on $\nu(r)$ for values up to 101; it is taken from [57]. A more extensive list appears in [52]. In 1987, Dinitz ([48]) proved:

Theorem 4.6 *If r is odd, $r \geq 17$, then $\nu(r) \geq 5$.*   □

The k-dimensional object equivalent to a k-POSLS$(2n - 1)$ is called a *Room k-cube* of side $2n - 1$; when $k = 3$, the term "Room cube" is used.

Exercises 4

4.1 Prove that a symmetric Latin square of odd side r must have each of the symbols $1, 2, \cdots, r$ exactly once on its main diagonal.

4.2 Find a pair of orthogonal Latin squares of side 3, and write down the corresponding pair of orthogonal one-factorizations of $K_{3,3}$.

4.3 Repeat Exercise 4.2 for side 4.

4.4 Prove that there is no one-factorization of $K_{4,4}$ orthogonal to $\{F_1, F_2, F_3, F_4\}$, where

$$
\begin{array}{llll}
F_1 & = & \{15, 26, 37, 48\} & F_2 & = & \{16, 27, 38, 45\} \\
F_3 & = & \{17, 28, 35, 46\} & F_4 & = & \{18, 25, 36, 47\}.
\end{array}
$$

4.5 Prove that there is no Room square of side 5.

4.6 Find a one-factorization of K_8 orthogonal to $\{G_1, G_2, G_3, G_4, G_5, G_6, G_7\}$, where

$$
\begin{array}{llll}
G_1 & = & \{01, 23, 45, 67\} & G_2 & = & \{02, 14, 36, 57\} \\
G_3 & = & \{03, 16, 25, 47\} & G_4 & = & \{04, 17, 26, 35\} \\
G_5 & = & \{05, 12, 37, 46\} & G_6 & = & \{06, 15, 27, 34\} \\
G_7 & = & \{07, 13, 24, 56\}.
\end{array}
$$

and write down the corresponding Room square.

4.7 Prove that there is no one-factorization of K_8 orthogonal to $\{H_1, H_2, H_3, H_4, H_5, H_6, H_7\}$, where

$$
\begin{aligned}
H_1 &= \{01, 23, 45, 67\}, & H_2 &= \{02, 13, 46, 57\} \\
H_3 &= \{03, 14, 27, 56\}, & H_4 &= \{04, 16, 25, 37\} \\
H_5 &= \{05, 17, 26, 34\}, & H_6 &= \{06, 14, 27, 35\} \\
H_7 &= \{07, 15, 24, 36\}.
\end{aligned}
$$

TOURNAMENT APPLICATIONS
OF ONE-FACTORIZATIONS

Suppose several baseball teams play against each other in a league. The competition can be represented by a graph with the teams as vertices and with an edge xy representing a game between teams x and y. We shall refer to such a league — where two participants meet in each game — as a *tournament*. (The word "tournament" is also used for the directed graphs derived from this model by directing the edge from winner to loser, but we shall not consider the results of games.)

Sometimes multiple edges will be necessary; sometimes two teams do not meet. The particular case where every pair of teams plays exactly once is called a *round robin tournament*, and the underlying graph is complete.

A very common situation is when several matches must be played simultaneously. In the extreme case, when every team must compete at once, the set of games held at the one time is called a *round*. Clearly the games that form a round form a one-factor in the underlying graph. If a round robin tournament for $2n$ teams is to be played in the minimum number of sessions, we require a one-factorization of K_{2n}, together with an ordering of the factors (this ordering is sometimes irrelevant). If there are $2n - 1$ teams, the relevant structure is a near-one-factorization of K_{2n-1}. In each case the (ordered) factorization is called the *schedule* of the tournament.

In many sports a team owns, or regularly plays in, one specific stadium or arena. We shall refer to this as the team's "home field". When the game is played at a team's home field, we refer to that team as the "home team" and the other as the "away team". Often the home team is at an advantage; and more importantly, the home team may receive a greater share of the admission fees. So it is usual for home and away teams to be designated in each match. We use the term *home-and-away schedule* (or just *schedule*) to refer to a round-robin tournament schedule in which one team in each game is labeled the home team and one the away team. Since this could be represented by orienting the edges in the one-factors, a home-and-away schedule is equivalent to an oriented one-factorization. It is very common to conduct a double round-robin, in which every team plays every other team twice. If the two matches for each pair of teams are arranged so that the home team in one is the away team in the other, we shall say the schedule and the corresponding oriented one-factorization of $2K_{2n}$ are *balanced*.

For various reasons one often prefers a schedule in which runs of successive away games and runs of successive home games do not occur (although there are exceptions: an east coast baseball team, for example, might want to make a tour of the west, and play several away games in succession). We shall define a *break* in a schedule to be a pair of successive rounds in which a given team is at home, or away, in both rounds. A schedule is *ideal* for a team if it contains no break for that team. Oriented factorizations are called ideal for a vertex if and only if the corresponding schedules are ideal for the corresponding team.

Theorem 5.1 [171] *Any schedule for $2n$ teams is ideal for at most two teams.*

Proof. For a given team x, define a vector v_x to have $v_{xj} = 1$ if x is home in round j and $v_{xj} = 0$ if x is away in round j. If the schedule is ideal for team x, then v_x consists of alternating zeroes and ones, so there are only two possible vectors v_x for such teams. But $v_x \neq v_y$ when $x \neq y$: the vectors must differ in position j, where x plays y in round j. So the schedule can be ideal for at most two teams.   □

The following theorem shows that the theoretical best-possible case can be attained.

Theorem 5.2 [171] *There is an oriented one-factorization of K_{2n} with exactly $2n - 2$ breaks.*

Proof. We orient the one-factorization $\mathcal{P} = \{P_1, P_2, \cdots, P_{2n-1}\}$ based on $\{\infty\} \cup \mathbb{Z}_{2n-1}$, defined by

$$P_k = \{(\infty, k)\} \cup \{(k + i, k - i) : 1 \leq i \leq n - 1\}. \qquad (5.1)$$

Edge (∞, k) is oriented with ∞ at home when k is even and k at home when k is odd. Edge $(k + i, k - i)$ is oriented with $k - i$ at home when i is even and $k + i$ at home when i is odd.

It is clear that ∞ has no breaks. For team x, where x is in $\mathbb{Z}_{2n-1}$, we can write $x = k + (x - k) = k - (k - x)$. The way in which x occurs in the representation (5.1) will be: as x when $k = x$, as $k + (x - k)$ when $1 \leq x - k \leq n - 1$, and as $k - (k - x)$ otherwise. The rounds other than P_x where x is at home are the rounds k where $x - k$ is odd and $1 \leq x - k \leq n - 1$, and the rounds k where $k - x$ is even and $1 \leq k - x \leq n - 1$. It is easy to check that factors P_{2j-1} and P_{2j} form a break for symbols $2j - 1$ and $2j$, and that these are the only breaks.

$\square$

Suppose two teams share the same home ground. (In practice this occurs quite frequently — the cost of maintaining a football stadium, for example, is high, and two teams will often share the expenses.) Then those two teams cannot both be "home" in the same round (except for the round in which they play each other). To handle the most demanding case, one needs a home-and-away schedule with the following property: there is a way of pairing the teams such that only one member of each pair is at home in each round. We shall refer to this as property (P). Schedules with this property are discussed in [17], where an example for four teams is given in order to construct a more complicated type of tournament.

Theorem 5.3 [160] *Given any one-factorization of K_{2n}, there exists a way of pairing the vertices and a way of orienting the edges such that the resulting home-and-away schedule has property (P) with respect to the given pairing.*

Proof. Select one factor at random; call it F_1. The pairs which play against each other in this round will constitute the pairing referred to in property (P). No matter how home teams are allocated in the round corresponding to F_1, only one member of each pair will be at home.

Now select any other round. Let the one-factor associated with this round be F_2. Consider the union of F_1 and F_2. In each of the cycles comprising it, select one of the points at random; the teams corresponding to it and to every second point as you go around the cycle from it are the home teams in the round under discussion.

Exactly one home team has been selected in every edge of F_2. So there is exactly one home team in every match of the new round. So (P) is satisfied. $\square$

Suppose a double round-robin tournament is to be played with the sort of home ground sharing which we have been using. This is easily managed when the tournament consists of two copies of an ordinary round-robin: first $2n - 1$ rounds are played with home teams allocated as in the proof of Theorem 5.3, and then the competition is repeated with home-and-away teams exchanged. However, not all double round-robins are of this type. The underlying factorization is a one-factorization of $2K_{2n}$, the multigraph on $2n$ vertices with exactly two edges joining each pair of distinct points. There are three types of factorization possible — two copies of a one-factorization of K_{2n}, copies of two different one-factorizations of K_{2n}, and factorizations which cannot be decomposed into two one-factorizations of K_{2n}. Factorizations of this third type are called *indecomposable*. We discuss indecomposable factorizations of complete multigraphs in Chapter 17.

Each of the three possibilities can occur, even in as small an example as $2K_6$; in fact, up to isomorphism, there is exactly one factorization of each kind in that case. They are shown in Table 5.1, in the order given above; factors are written as columns. Each factorization has been written so that, if the left-hand team in each match is the home team, then the pairing 1-2, 3-4, 5-6 satisfies (P), and each pair xy occurs once in a game with x at home and once with y. Let us call this a *proper orientation* of the factorization with regard to the pairing; Table 5.1 shows that every one-factorization of $2K_6$ has a proper orientation. In general there is no known example of a one-factorization of $2K_{2n}$ with no proper orientation, although it has not been shown that none exists, even in the case of $2K_8$.

12	13	14	15	16	21	31	41	51	61
34	52	62	42	32	43	25	26	24	23
56	46	35	63	54	65	64	53	36	45
12	13	14	51	61	21	31	41	15	16
34	52	62	24	23	43	26	25	32	42
56	46	35	36	45	65	54	63	64	53
12	21	31	13	14	41	51	15	16	61
34	35	24	62	52	26	23	42	32	25
56	64	65	45	36	53	46	63	54	43

Table 5.1 The one-factorizations of $2K_6$, properly oriented.

A number of other problems have been studied concerning scheduling and concerning the sharing of facilities; one particular problem concerns a competition with a junior and a senior league, where many clubs field teams in both leagues; naturally, teams from the same club share a stadium. The interested reader should consult, for example, [17, 145, 172, 173].

Problems can arise in the allocation of games even without home-and-away considerations. Suppose one wishes to run a round robin tournament with $2n$ teams. One technique is to play round 1 by pairing teams at random. The matches in round 2 are also arbitrary, but one demands "compatibility" — no two teams play if they already met in round 1. If this process is continued, will it always be successful or are there cases where this premature assignment of matches, without checking to see whether the tournament can be completed, will result in an impossible situation? In fact, such unsatisfactory results can occur. The smallest case occurs when $2n = 6$: the rounds

(1,4)	(2,5)	(3,6)
(1,5)	(2,6)	(3,4)
(1,6)	(2,4)	(3,5)

cannot be completed. We discuss such problems in Chapter 20, and related problems in Chapters 18 and 19.

In some sports there is a carryover effect from round to round: for example, if team X plays against a very strong team, then X may perform poorly in its next round, either because its members are demoralized or because it was weakened in the preceding round. To overcome this bias, Russell [133] proposed the idea of a tournament which is balanced for carryover.

Consider a tournament whose schedule is the ordered one-factorization $\mathcal{F}$,

$$\mathcal{F} = \{F_0, F_1, \cdots, F_{2n-2}\},$$

of the K_{2n} with vertex-set S. It will be convenient to write x_i for the team which opposes x in round i, so that

$$F_i = \{(x, x_i) : x \in S\}.$$

(Of course, each edge is written twice in this representation.) We define the *predecessor* of y at round j to be that x such that

$$x_j = y_{j+1},$$

and denote it by $p_j(y)$; $x = p_{2n-2}(y)$ means $x_{2n-2} = y_1$. The set of predecessors of y is $P(y)$,

$$P(y) = \{p_j(y) : 0 \leq j \leq 2n - 2\}.$$

$\mathcal{F}$ is *balanced for carryover* if

$$P(y) = S\backslash\{y\} \text{ for all } y \in S.$$

Theorem 5.4 [133] *There is a tournament schedule which is balanced for carryover whenever the number $2n$ of teams is a power of 2.*

Proof. (after [92]) Say $2n = 2^m$, for some positive integer m. Select a primitive element ξ in the field $GF(2n)$. We take $S = GF(2n)$, so

$$S = \{0, 1, \xi, \cdots, \xi^{2n-2}\}.$$

We define the factor F_0 to consist of all the pairs $(z, z + 1)$ where $z \in S$, and form F_i by multiplying every member of F_0 by ξ^i:

$$F_i = \{(\xi^i z, \xi^i(z + 1)) : z \in S\}.$$

If $x = \xi^i z$ then

$$x_i = \xi^i(z + 1) = x + \xi^i.$$

Say $p_j(y) = x$. Then

$$\begin{aligned} x_j &= y_{j+1}, \\ x + \xi^j &= y + \xi^{j+1}, \\ x &= y + \xi^j(1 + \xi). \end{aligned}$$

So

$$P(y) = \{y + \xi^j(1 + \xi) : 0 \leq j \leq 2n - 2\}.$$

As j varies from 0 to $2n - 2$, ξ^j ranges through the non-zero elements of S, and so does $\xi^j(1 + \xi)$. (Certainly $1 + \xi$ is non-zero, as otherwise $\xi = 1$, which is not primitive). So

$$\begin{aligned} \{\xi^j(1 + \xi)\} &= S\backslash\{0\}, \\ P(y) &= S\backslash\{y\}, \end{aligned}$$

and $\mathcal{F}$ is balanced for carryover. $\square$

As an example, the schedule for eight teams is shown in Table 5.2. (The power ξ^i is represented by the integer $i + 1$.) A complete search shows that there is no

01	26	34	57
02	37	45	61
03	41	56	72
04	52	67	13
05	63	71	24
06	74	12	35
07	15	23	46

Table 5.2   A schedule which is balanced for carryover.

example of order 6 or 10, and the only examples of order 8 are the one shown and the one derived from it by reversing the order of the factors. Russell [133] has conjectured that schedules balanced for carryover exist only when the order is a power of 2.

The construction just given is a special case of the following idea. Consider orderings of the field $GF(q)$. An ordering $(0 = a_0, a_1, \cdots, a_{q-1})$ is called *balanced* if the partial sums $S_1, S_2, \cdots, S_{q-1}$ are all different, where

$$S_t = \sum_{i=1}^{t} a_i.$$

Given a balanced ordering of $GF(q)$, write a_q for the (unique) element of the field which does not arise as one of $\{S_1, S_2, \cdots, S_{q-1}\}$. Define

$$b_{i,r} = a_i + a_q + S_{r-1}. \tag{5.2}$$

Theorem 5.5 [133]   If $(a_0, a_1, \cdots, a_{q-1})$ is a balanced ordering of $GF(q)$, where $q = 2^m$, then

$$F_r = \{\{a_i, b_{i,r}\}\}$$

(where $b_{i,r}$ is defined as in (5.2)) is a one-factor of K_q, and

$$\{F_1, F_2, \cdots, F_{q-1}\}$$

is a one-factorization and is balanced for carryover. $\square$

Verifying this Theorem is left to the reader. Notice that if ξ is a primitive element in $GF(q)$ then $(0, 1, \xi, \xi^2, \cdots)$ is a balanced ordering — if $S_i = S_j$ when $1 \leq i < j \leq q - 1$ then

$$\xi^i + \xi^{i+1} + \cdots + \xi^{j-1} = 0,$$

so

$$1 + \xi + \cdots + \xi^{j-i-1} = 0;$$

multiplying both sides by $1 - \xi$ yields $\xi^{j-i} = 1$, which is impossible. So Theorem 5.4 is a special case of Theorem 5.5.

A further restriction on schedules was proposed in [19]. We say that x is the *ℓ-th predecessor* of y at round j, and write $x = p_{j,\ell}(y)$, if $x_j = y_{\ell+j}$ (addition of subscripts being carried out modulo $2n - 1$). Write

$$P_\ell(y) = \{p_{j,\ell}(y) : 0 \leq j \leq 2n - 2\}.$$

Then the schedule is *balanced at level ℓ* if $P_\ell(y) = S \backslash \{y\}$, and it is *totally balanced* if it is balanced at every level.

It is easy to check that the construction of Theorem 5.4 is totally balanced. However, not all balanced orderings of $GF(2^m)$ yield totally balanced factorizations. Bonn [24] has generated all balanced orderings of $GF(16)$. He found many which are balanced at levels 1 and 14 only, one (and its reverse) which is totally balanced, and two different tournaments which are balanced at levels 1, 3, 6, 9, 12 and 14 but no others. The reason for the latter pattern is a mystery.

It is not difficult to prove

Theorem 5.6 [19] *If there exists a totally balanced tournament of order $2n$, then there exists a set of $n - 1$ pairwise orthogonal Latin squares of order $2n$.*

□

This Theorem gives some weight to a weaker version of Russell's conjecture: that a totally balanced schedule of order $2n$ can exist only when $2n$ is a power of 2.

Finally, one can simultaneously consider balance with respect to several different factors. Finitzio [65] has recently studied some problems of this kind and presents some general results as well as several interesting particular solutions.

Exercises 5

5.1 Two chess clubs, each of n members, wish to play a match over n nights. Each player will play one game per night against a different opponent from the other team. What mathematical structure is used? Give an example for $n = 4$.

5.2 Suppose that in Exercise 5.1 we interpret playing white as a home field advantage. Prove that there is a tournament schedule which may be oriented so that no player has a break (two successive plays as the same color).

5.3 Prove that there is a schedule for $2n - 1$ teams which is ideal for every team, in the sense that no team plays two home or two away games in consecutive rounds.

5.4 Prove the existence of a double round robin in which home matches are allocated in such a way that two teams have no breaks and every other team has exactly one break.

5.5 (*) A softball league plays in two conferences, each of four teams. Games are scheduled on Monday, Tuesday, Thursday and Friday nights, two games per night, plus an opening week of one game per night. No team may play twice per night, nor on consecutive nights. Teams meet other teams in the same conference three times and teams in the other conference twice. Construct a suitable schedule. [145]

5.6 (?) Is there a one-factorization of $2K_{2n}$, for any n, which has no proper orientation for any pairing?

5.7 Suppose F_1, F_2 and F_3 are three edge-disjoint one- factors of $2K_6$. Show that there is a pair $\{x, y\}$ such that x precedes y at round 1 and also at round 2. (Consequently there is no schedule for six teams which is balanced for carryover.)

5.8 (?) Is there any schedule for $2n$ teams which is balanced for carryover in any case when $2n$ is not a power of 2?

5.9 Prove Theorem 5.5.

5.10 Prove that if a one-factorization of K_{2n} is balanced at level i, then it is balanced at level $2n - 1 - i$.

5.11   (i) Suppose the schedule $\mathcal{F}$ is balanced for carryover. Define an array
       $A^0 = (a_{i,j})$ of order $2n$ by

$$a_{i,j} = j_i, \ 0 \leq i \leq 2n - 2,$$
$$a_{2n-1,j} = j.$$

Prove that A^0 is a Latin square.

(ii) If $1 \leq k \leq 2n - 2$, define $A^k = (a_{i,j}^k)$ by

$$a_{i,j}^k = (j_i)_{i-k}, \ 0 \leq i \leq 2n - 2,$$
$$a_{2n-1,j}^k = j.$$

Prove that if $\mathcal{F}$ is totally balanced then $A^1, A^2, \cdots, A^{2n-2}$ are Latin squares.

(iii) Prove Theorem 5.6.

6

A GENERAL EXISTENCE
THEOREM

If W is any subset of the vertex-set $V(G)$ of a graph or multigraph G, we write $G - W$ to denote the graph constructed by deleting from G all vertices in W and all edges touching them. One can discuss the components of $G - W$; they are either *odd* (have an odd number of vertices) or *even*. Let $\Phi_G(W)$ denote the number of odd components of $G - W$.

Theorem 6.1 [150] *G contains a one-factor if and only if*

$$\Phi_G(W) \le |W| \text{ whenever } W \subset V(G). \tag{6.1}$$

Proof. (after [109]) We prove the Theorem on the assumption that G is a graph. If multiple edges are allowed in G, write G' for the underlying graph of G. The truth of the Theorem for G' implies its truth for G.

First suppose G contains a one-factor F. Select a subset W of $V(G)$, and suppose $\Phi_G(W) = k$; label the odd components of $G - W$ as $G_1, G_2, \cdots, G_k$. As G_i has an odd number of vertices, $G_i \backslash F$ cannot consist of $\frac{1}{2}|G_i|$ edges; there must be at least one vertex, x say, of G_i which is joined by F to a vertex y_i which is not in G_i. Since components are connected, y_i must be in W. So W contains at least the k vertices $y_1, y_2, \cdots, y_k$, and

$$k = \Phi_G(W) \le |W|.$$

So the condition is necessary.

To prove sufficiency, we assume the existence of a graph G which satisfies (6.1) but has no one-factor; a contradiction will be obtained. If such a G exists,

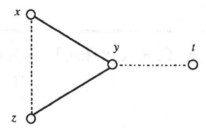

Figure 6.1 A subgraph arising in the proof of Theorem 6.1.

we could continue to add edges until we reached a maximal graph G^* such
that no further edge could be added without introducing a one-factor. (Such a
maximum exists: it follows from the case $W = \emptyset$ that G has an even number of
vertices; if we could add edges indefinitely, eventually an even-order complete
graph would be reached.) Moreover the graph G^* also satisfies (6.1) — adding
edges may reduce the number of odd components, but it cannot increase them.
So there is no loss of generality in assuming that G is already maximal. We
write U for the set of all vertices of degree $|V(G)| - 1$ in G. If $U = V(G)$ then G
is complete, and has a one-factor. So $U \neq V(G)$. Every member of U is adjace
nt to every other vertex of G. We first show that every component of $G - U$
is a complete graph. Let G_1 be a component of $G - U$ which is not complete.
Not all vertices of G_1 are joined, but G_1 is connected, so there must exist two
vertices at distance 2 in G_1, x and z say; let y be a vertexadjacent to both.
Since $d(y) \neq |V(G)| - 1$ there must be some vertex t of G which is notadjacent
to y, and t is in $V(G - U)$ (since every vertex, y included, isadjacent to every
member of U). So $G - U$ contains the configuration shown in Figure 6.1 (a
dotted line means "no edge").

Since G is maximal, $G + xz$ has a one-factor — F_1 say — and $G + yt$ also has
a one-factor — call it F_2. Clearly xz belongs to F_1 and yt belongs to F_2; and
also xz is not in F_2.

Consider the graph $F_1 \cup F_2$; let H be the component that contains xz. As xz is
not in F_2, H is a cycle of even length made up of alternate edges from F_1 and
F_2. Either yt belongs to H or else yt belongs to another even cycle of alternate
edges. These two cases are illustrated in Figures 6.2(a) and 6.2(b).

In case (a) we can assume that vertices x, y, t and z appear in that order along
the cycle H, as shown in the Figure (if not, interchange x and z). Then we

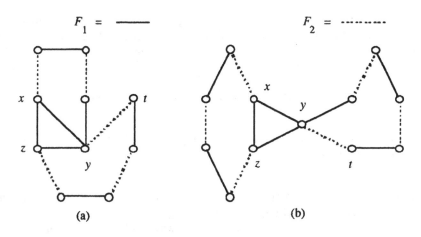

Figure 6.2 Two cases needed in Theorem 6.1.

can construct a one-factor of G as follows: take the edges of F_1 in the section $y, t, \cdots, z$ of H, the edges of F_2 in the rest of G, and yz. In case (b) we can construct a one-factor in G by taking the edges of F_1 in H and the edges of F_2 in $G - H$. So in both cases we have a contradiction. So G_1 cannot exist — every component of $G - U$ must be complete.

Now $\Phi_G(U) \leq |U|$, so $G - U$ has at most $|U|$ odd components. We associate with each odd component G_i of $G - U$ a different member u_i of U. We then construct a one-factor of G as follows. From every even component of $G - U$ select a one-factor (possible because the components are complete graphs); for each odd component G_i, select a one-factor of $G_i + u_i$ (again, $G_i + u_i$ is an even-order complete graph). Since G has an even number of vertices, there will be an even number of vertices left over, all in U; as they are all connected, a one-factor can be chosen from them. The totality is a one-factor in G, contradicting the hypothesis. $\square$

Suppose G has v vertices and suppose the deletion of the w-set W of vertices results in a graph with t odd components; then $v - w \equiv t \pmod{2}$. If v is even then $w \equiv t \pmod{2}$, and $t > w$ will imply $t \geq w + 2$. So we have a slight improvement on Theorem 6.1.

Theorem 6.2 [156] *If the graph or multigraph G has an even number of vertices, then G has no one-factor if and only if there is some w-set W of vertices such that $G - W$ has at most $w + 2$ odd components.*   □

As an application of Theorem 6.1 we prove the following result.

Theorem 6.3 [42] *If n is even then any regular graph of degree $n - 1$ on $2n$ vertices has a one-factor.*

Proof. If $n = 2$ or $n = 4$ then the result is easily checked by considering all cases. So we assume G is a regular graph of degree $n - 1$ on $2n$ vertices, where n is even and $n > 4$, and W is any set of w vertices of G, and we prove that the graph $G - W$ has at most w odd components.

If $w \geq n$, then $G - W$ has at most w vertices, so it has at most w components. If $w = 0$, then G has an odd component, which is impossible since w has odd degree. So we assume $1 \leq w \leq n - 1$.

The deletion of W cannot reduce the degree of a vertex of G by more than w, so every vertex of $G - W$ has degree at least $n - 1 - w$, and each component has at least $n - w$ vertices. If there are $w + 1$ or more components then $G - W$ has at least $(w + 1)(n - w)$ vertices, and

$$(w + 1)(n - w) \leq 2n - w,$$

which simplifies to

$$w^2 - wn + n \geq 0.$$

For fixed n this is a quadratic inequality which will have the solution

$$w \leq w_1 \text{ or } w \geq w_2$$

where w_1 and w_2 are the roots of

$$w^2 - wn + n = 0 :$$

that is,

$$w_1 = \tfrac{1}{2}(n - \sqrt{n^2 - 4n}), \quad w_2 = \tfrac{1}{2}(n + \sqrt{n^2 - 4n}).$$

Now when $n > 4$,

$$n - 3 < \sqrt{n^2 - 4n} < n - 2,$$

so
$$1 < w_1 < 2, \quad n - 2 < w_2 < n - 1,$$

and the only integer values in the range $1 \le w \le n - 1$ which satisfy the inequality are $w = 1$ and $w = n - 1$.

If $w = 1$ then $G - W$ has every vertex of degree at least $n-2$, so every component has at least $n - 1$ vertices. The only possible case is two components, one with $n - 1$ vertices and one with n. Only one is odd, so $G - W$ has at most $w \, (= 1)$ odd components.

If $w = n - 1$ then $G - W$ has $n + 1$ vertices. To get $w + 1$ odd components the only possibility is $n + 1$ components, each of one vertex. So $G - W$ is empty: deletion of W has removed all $n(n - 1)$ edges of G. Since each vertex of W had degree $n - 1$, at most $(n - 1)^2$ edges can have been removed, which is a contradiction. $\qquad \Box$

As another application we shall prove Petersen's theorem which we promised in Chapter 3. In fact we prove a more general result which has Petersen's Theorem as a corollary.

If W is any set of vertices of a graph G, define $z_G(W)$ to be the number of edges of G with precisely one endpoint in W.

Lemma 6.4 *If G is a regular graph or multigraph of degree d and S is any set of vertices of G, then*
$$d|S| = 2e(\langle S \rangle) + z_G(S). \qquad (6.2)$$

Proof. The sum of the degrees of vertices in S, and $2e(\langle S \rangle)$ is the contribution from edges with both endpoints in S; (6.2) follows. $\qquad \Box$

Theorem 6.5 *If G is a regular graph or multigraph of degree d with an even number of vertices, and if $z_G(S) \ge d - 1$ for every odd-order subset S of $V(G)$, then each edge of G belongs to some one-factor.* $\qquad \Box$

Proof. Consider the edge xy of G. We prove that $H = G - \{x, y\}$ has a one-factor. Then the union of this factor with $\{xy\}$ is the required one-factor.

If equation (6.2) is reduced modulo 2, the result is $d \equiv z_G(S)(\text{mod } 2)$ when $|S|$ is odd, so the data of the Theorem actually imply that $z_G(S) \geq d$ for every odd-order set S of vertices.

Suppose H has no one-factor. From Theorem 6.2, $\Phi_H(X) \geq |X| + 2$ for some subset X of $V(H)$. So $\Phi_G(W) \geq |W|$ where W is the subset $X \cup \{x, y\}$ of $V(G)$. If S is any odd component of $G - W$, every edge joining S to $G - S$ must in fact have an endpoint in W, so there are $z_G(S)$ such edges. We observe that the data of the Theorem actually imply that $z_G(S) \geq d$ for every odd-order set S of vertices, because equation (6.2) means that $d|S$ and $z_G(S)$ are congruent modulo 2, so there are at least d edges joining S to W. Summing over all odd components of $G - W$,

$$z_G(W) \geq d\Phi_G(W) \geq d|W|.$$

But $\langle W \rangle$ contains at least one edge, xy, so from (6.2)

$$z_G(W) \leq d|W| - 2,$$

— a contradiction.   □

When $d = 3$, the condition $z_G(W) \geq d - 1$ is equivalent to saying that no edge incident with W is a bridge, so:

Corollary 6.5.1 *If G is a bridgeless cubic graph or multigraph and e is any edge G, then G has a one-factor which contains e.*   □

In particular this implies Petersen's Theorem, as promised.

Corollary 6.5.2 [44] *If G is a regular graph with $2m$ vertices such that $d \geq m$ (m odd) or $d \geq m - 1$ (m even), then every edge of G belongs to some one-factor.*
                                                                            □

Theorem 6.5 was discovered by Berge [20], and later proven by Cruse [44] found another demonstration using a Theorem on decomposing stochastic matrices. The easy and elegant proof above is due to West.

Cruse went on to prove the following Theorem.

Theorem 6.6 [44] *If G is a regular graph of degree d with $2m$ vertices, and $d \geq m \geq 2$, then G contains at least*

$$d - m + 2 \left\lceil \frac{5(m+5)}{112} \right\rceil$$

disjoint one-factors. $\square$

Exercises 6

6.1 Does Theorem 6.3 apply to multigraphs?

6.2 Prove that there are exactly three regular graphs of degree 3 on eight vertices (up to isomorphism), and that each has a one-factor.

6.3 Prove that if a cubic graph has at most two bridges then it has a one-factor. [134]

6.4 Suppose G is a regular graph of degree d which has a one-factorization. Prove that $z_G(W) \geq d - 1$ for every odd-order subset W of $V(G)$. Is this necessary condition sufficient?

6.5 Prove Corollary 6.5.2.

6.6 G_1 and G_2 are complete graphs on $2k + 1$ vertices. G is derived from $G_1 \cup G_2$ by deleting an edge $u_1 v_1$ from G_1 and an edge $u_2 v_2$ from G_2 and adding edges $u_1 u_2$ and $v_1 v_2$. Prove that each edge of G belongs to some one-factor, but G does not satisfy the conditions of Theorem 6.5.

6.7 Prove that if a graph is regular of degree 5 and it has edge-connectivity at least 4 then it has a one-factor.

6.8 G is n-connected, regular of degree n, and has an even number of vertices. Prove that G has a one-factor.

6.9 Prove that a graph G has a one-factor if and only if $G - v$ contains precisely one odd component for each vertex v of G.

6.10 A tree T contains a one-factor. Prove that the factor is uniquely defined.

7

GRAPHS WITHOUT ONE-FACTORS

Suppose G is a regular graph of degree d and suppose $G - W$ has a component with p vertices, where p is no greater than d. The number of edges within the component is at most $\frac{p(p-1)}{2}$. This means that the sum of the degrees of these p vertices in $G - W$ is at most $p(p-1)$. But in G each vertex has degree d, so the sum of the degrees of the p vertices is pd, whence the number of edges joining the component to W must be at least $pd - p(p-1)$. For fixed d and for integer p satisfying $1 \le p \le d$ this function has minimum value d (achieved at $p = 1$ and $p = d$). So any odd component with d or fewer vertices is joined to W by d or more edges.

We now assume that G is a regular graph of degree d on v vertices, where v is even. G has no one-factor; by Theorem 6.2 there is a set W of w vertices whose deletion leaves at least $w + 2$ odd components. We call a component of $G - W$ *large* if it has more than d vertices, and *small* otherwise. The numbers of large and small components of $G - W$ are α_W and β_W, or simply α and β, respectively. Clearly

$$\alpha + \beta \ge w + 2. \tag{7.1}$$

There are at least d edges of G joining each small component of $G - W$ to W, and at least one per large component, so there are at least $\alpha + d\beta$ edges attached to the vertices of W; by regularity we have

$$\alpha + d\beta \le wd. \tag{7.2}$$

Each large component has at least $d + 1$ vertices if d is even, and at least $d + 2$ if v is odd, so

$$
\begin{aligned}
v &\ge w + (d+1)\alpha + \beta \text{ if } d \text{ is even,} & (7.3) \\
v &\ge w + (d+2)\alpha + \beta \text{ if } d \text{ is odd,} & (7.4)
\end{aligned}
$$

51

Since α is non-negative (7.2) yields $\beta \leq w$, so from (7.1) we have $\alpha \geq 2$; but applying this to (7.2) again we get $\beta < w$, so from (7.1) we have $\alpha \geq 3$. So from (7.3) and (7.4) we see that if $w \geq 1$ then

$$v \;\geq\; 3d + 4 \text{ if } d \text{ is even}, \tag{7.5}$$
$$v \;\geq\; 3d + 7 \text{ if } d \text{ is odd}, \tag{7.6}$$

In the particular case $d = 4$, the bound in (7.5) cannot be attained. For, suppose P is an odd component of $G - W$ with p vertices. Then the sum of the vertices in G of members of P is $4p$, which is even. On the other hand, if there are r edges joining W to P in G and s edges internal to W, the sum of the degrees is $r + 2s$. So r is even. So there are at least 2 edges from each large component to W, and 7.2 can be strengthened to

$$2\alpha + 4\beta \leq 4t,$$

whence $2\beta \leq 4k - (2\alpha + 2\beta)$; substituting from (7.1) we get

$$2\beta \leq 2k - 4$$

and $\alpha \geq 4$. So (7.3) yields

$$v \geq 2 + 5 \cdot 4 = 22.$$

Summarizing this discussion, we have:

Theorem 7.1 [156] *If a regular graph G with an even number v of vertices and with degree d has no one-factor and no odd component, then*

$$v \;\geq\; 3d + 7 \text{ if } d \text{ is odd}, \ d \geq 3;$$
$$v \;\geq\; 3d + 4 \text{ if } d \text{ is even}, \ d \geq 6;$$
$$v \;\geq\; 22 \text{ if } d = 4. \qquad \qquad \square$$

The condition of "no odd component" is equivalent to the assumption that $w \geq 1$. The cases $d = 1$ and $d = 2$ are omitted, but in fact every graph with these degrees which satisfies the conditions has a one-factorization.

It follows from the next result that Theorem 7.1 is best-possible.

Theorem 7.2 *If v is even and is at least as large as the bound of Theorem 7.1, then there is a regular graph of the relevant degree on v vertices which has no one-factor.*

Proof. We use two families of graphs. The graph $G_1(h, k, s)$ has $2s+1$ vertices, and is formed from K_{2s+1} as follows. First factor K_{2s+1} into Hamilton cycles, as in Theorem 3.3. Then take the union of $h - 1$ of those cycles. Finally take another of the cycles, delete k disjoint edges from it, and adjoin the remaining edges to the union. This construction is possible whenever $0 < h \leq s$ and $0 \leq k \leq s$. If a graph has $2s + 1$ vertices, of which $2k$ have degree $2h - 1$ and the rest have degree $2h$, let us call it a $1 - (h, k, s)$ graph; for our purposes, the essential property of $G_1(h, k, s)$ is that it is a $1 - (h, k, s)$ graph.

The graph $G_2(h, k, s)$ has $2s+1$ vertices. We construct it by taking a Hamilton cycle decomposition of K_{2s+1}, and taking the union of h of the cycles. Then another cycle is chosen from the factorization; from it are deleted a path with $2k + 1$ vertices and $s - k$ edges which contain each of the remaining $2s - 2k$ vertices once each. We define a $2 - (h, k, s)$ graph to be a graph on $2s + 1$ vertices with $2k - 1$ vertices of degree $2h$ and the rest of degree $2h + 1$. Then $G_2(h, k, s)$ is a $2 - (h, k, s)$ graph whenever $0 < h < s$ and $0 < k \leq s$.

Finally we define the composition $[G, H, J]$ of three graphs G, H and J, each of which have some vertices of degree d and all other vertices of degree $d - 1$, to be the graph formed by taking the disjoint union of G, H and J and adding to it a vertex x and edges joining all the vertices of degree $d - 1$ in G, H and J to x. It is clear that if d is even, say $d = 2h$, and $h \geq 3$, then

$$[G_1(h, 1, h), G_1(h, 1, h), G_1(h, h - 2, h + t)]$$

is a connected regular graph of degree d on $3d + 4 + 2t$ vertices which has no one-factor, since the deletion of x results in three odd components. Similarly if $d = 2h + 1$ then

$$[G_2(h, 1, h + 1), G_2(h, 1, h + 1), G_2(h, h, h + t + 1)]$$

is a connected regular graph of degree d on $3d + 7 + 2t$ vertices which has no one-factor.

The first construction does not give a connected graph when $h = 2$, so another construction is needed for the case $d = 4$. Take three copies of $G_1(2, 1, 2)$, which each have five vertices, two of degree 3 and three of degree 4, and one copy of $G_1(2, 1, 2 + t)$, which has two vertices of degree 3 and $3 + 2t$ of degree 4. Add two new vertices, x and y; join one vertex of degree 3 from each of the four graphs to x and the other to y. The result has $22 + 2t$ vertices, degree 4 and no one-factor. $\qquad\square$

Pila [127] has considered the extension of this problem to the case of graphs with specified connectivity. His result is as follows.

Theorem 7.3 *Suppose G is a regular graph of degree d and connectivity k on v vertices, where v is even; and suppose G has no one-factor. There is no such graph if $d \leq 2$; and:*

$$
\begin{aligned}
\text{if } k &= 2 \text{ and } d \text{ is even then } v \geq 4d + 6; \\
\text{if } k &= 2 \text{ and } d = 5 \text{ then } v \geq 38; \\
\text{if } k &= 2 \text{ and } d = 7 \text{ then } v \geq 38; \\
\text{if } k &= 2 \text{ and } d \text{ is odd }, d \geq 9, \text{ then } v \geq 4d + 8; \\
\text{if } k &= 3 \text{ and } d = 7 \text{ then } v \geq 46; \\
\text{if } k &= 3 \text{ and } d = 8 \text{ then } v \geq 48; \\
\text{if } k &= 3 \text{ and } d = 10 \text{ then } v \geq 58; \\
\text{if } k &= 4 \text{ and } d = 9 \text{ then } v \geq 68;
\end{aligned}
$$

while if $d - k = 2$ or 3, then

$$
\begin{aligned}
v &\geq d^2 + 2d - 2, \ d \text{ even}, \\
v &\geq d^2 + 3d - 2, \ d \text{ odd};
\end{aligned}
$$

and for other cases when $k \geq 3$,

$$
v \geq nD + 2d + 2,
$$

where

$$
\begin{aligned}
D &= d \text{ if } d \text{ is even}, \\
&= d + 1 \text{ if } d \text{ is odd},
\end{aligned}
$$

and

$$
\begin{aligned}
n &= \left\lceil \frac{2d}{d - k} \right\rceil \text{ if } d - k \text{ is even}, \\
n &= \left\lceil \frac{2d}{d - k - 1} \right\rceil \text{ if } d - k \text{ is odd}.
\end{aligned}
$$

Moreover these results are best-possible, in the sense that there is a graph without a one-factor in each case where equality is assumed. □

The proof of Theorem 7.3 is quite lengthy, as it involves a number of special cases, but is not essentially different in style to the cases we have already studied.

Katerinis [98] has achieved part of Pila's results by considering the maximum size of a set of independent edges.

A different generalization was suggested by a question in [166]. A graph is *almost regular* if all vertices except one have the same degree. Let us define an $A(2n, d, e)$-graph to be an almost regular graph with $2n$ vertices, one of degree $d + e$ and the others of degree d, which has no one-factor. For what orders $2n$ does such a graph exist? Mardiyono [31, 113] proved the following result.

Theorem 7.4 *There is an $A(2n, d, e)$-graph when e is even and:*

(i) $e \geq 4, d = 2$ *and* $e \geq 6$;

(ii) d *is odd,* $d \geq 3$ *and*

$$2n \geq \begin{cases} e + d + 1, & \text{if } e \geq 2d, \\ 3d + 3, & \text{if } d + 1 \leq e \leq 2d - 2, \\ 3d + 5, & \text{otherwise}; \end{cases}$$

(iii) d *is even,* $d \geq 4$ *and*

$$2n \geq \begin{cases} e + d + 2, & \text{if } e \geq 3d + 4, \\ e + d + 4, & \text{if } 2d \leq e \leq 3d + 2, \\ 3d + 4, & \text{otherwise}. \end{cases} \qquad \square$$

The details are left as an exercise.

Exercises 7

7.1 Prove that $G_1(h, k, h)$ is uniquely defined up to isomorphism and that it is the only $1 - (h, k, h)$ graph. Is the corresponding result true of $2 - (h, k, h + 1)$-graphs?

7.2 What is the smallest value s such that there is a $1 - (h, 1, s)$ graph not of the type $G_1(h, 1, s)$?

7.3 (?) Is the following conjecture true: given v, k and d such that v exceeds or equals the bound in Theorem 7.3, then there is a regular graph on v vertices which has connectivity k, degree d and no one-factor? (Theorem 7.3 covers the "equals" case.)

7.4 (*) Suppose G is a k-connected regular graph of degree d, where $d > k \geq 1$, of even order v. Prove that G has a one-factor if:

   (i) $k = 1$ and $v \leq 6\lceil \frac{d}{2} \rceil + 3$;

   (ii) $k = 2$ and $v \leq 8\lceil \frac{d}{2} \rceil + 3$;

   (iii) $k \geq 3$ and $v \leq 10\lceil \frac{d}{2} \rceil + 1$.

7.5 (*) Prove Theorem 7.4.

EDGE-COLORINGS

If a graph is not regular, it cannot have a one-factorization. Edge-colorings provide the obvious generalization. A *k-edge-coloring* π of a graph G is a map from $E(G)$ to $\{1, 2, \cdots, k\}$, with the property that if e and f are edges with a common vertex then $\pi(e) \neq \pi(f)$. G is k-edge-colorable if there is a k-edge-coloring of G; the edge-chromatic number $\chi'(G)$ of G is the smallest number k such that G is k-edge-colorable. $\chi'(G)$ is also called the *chromatic index* of G. Given an edge-coloring π of G, we define the *spectrum* $S_\pi(x)$ (or simply $S(x)$) to be the set of size $d(x)$ defined by

$$S_\pi(x) = \{i : \pi(xy) = i \text{ for some } y \sim x\}.$$

The *color classes* under π are the k sets

$$C_i = C_i(\pi) = \{e : e \in E(G), \pi(e) = i\}.$$

Although C_i is defined as a set of edges, there is no confusion if it is also interpreted as the spanning subgraph with that edge-set.

It is clear that it requires $d(x)$ colors to color the edges at x. Recall that $\Delta(G)$ denotes the maximum of the degrees of vertices in G: so $\chi'(G) \geq \Delta(G)$. However we can say rather more. To do this we introduce a slightly more general type of coloring. A *k-painting* of G is a way of allocating k colors to the edges of G. We extend the preceding notations to k-paintings. If π is a k-painting but not necessarily an edge-coloring, and $s(x)$ (or $s_\pi(x)$) denotes the order of $S_\pi(x)$, then we can only say $s(x) \leq d(x)$, because there might be more than one edge in a given color touching x; a painting is an edge-coloring if and only if $s(x) = d(x)$ for all x. The *order* $|\pi|$ of a painting π equals the sum $\sum s(x)$ over all vertices x; π is a *maximal k-painting* if $|\pi|$ equals the maximum for the value of k.

Lemma 8.1 *If G is a connected graph other than an odd cycle, then there is a 2-painting of G in which $s(x) = 2$ whenever $d(x) \geq 2$.*

Proof. First, suppose G is Eulerian. If G is an even cycle, the obvious one-factorization provides the appropriate painting. Otherwise G has at least one vertex x of degree 4 or more. Select an Euler walk $e_1, e_2, \cdots, e_m$ where x is the start-finish vertex. Then the painting

$$C_1 = \{e_1, e_3, \cdots\}, \ C_2 = \{e_2, e_4, \cdots\}$$

has the required property, because every vertex (including x) is internal to the walk.

If G is not Eulerian, construct a new graph H by adding a vertex y to G and adding an edge xy whenever x is of odd degree. Then H is Eulerian, by Theorem 2.4. Carry out a painting of H using the Euler walk, and restrict to G: again the conditions are satisfied.   □

Lemma 8.2 *Suppose π is a maximal k-painting of G and suppose vertex x lies on two edges of color 1 and no edges of color 2. Then the component of $C_1 \cup C_2$ which contains x is an odd cycle.*

Proof. Suppose $C_1 \cup C_2$ is not an odd cycle. Then by Lemma 8.1 it has a 2-coloring ρ in which every vertex of degree at least 2 lies on edges of both colors. Replace C_1 and C_2 in π by the two color-classes of ρ, and denote the new k-coloring by σ. Then

$$s_\sigma(x) = s_\pi(x) + 1$$

and for every $y \neq x$

$$s_\sigma(y) \geq s_\pi(y).$$

So $|\sigma| > |\pi|$, which is a contradiction.   □

Theorem 8.3 *If G is a bipartite graph, then*

$$\chi'(G) = \Delta(G).$$

Proof. Suppose G is a graph for which $\chi'(G) > \Delta(G)$. Let π be a maximal k-painting of G, and select a vertex x for which $s(x) < d(x)$. Then x must satisfy the conditions of Lemma 8.2. Consequently G contains an odd cycle, and is not bipartite.   □

Corollary 8.3.1 *Every regular bipartite graph has a one-factorization.*   □

The following Theorem is due to Vizing [152] and was independently discovered by Gupta [80]. Our proof follows Fournier [68].

Theorem 8.4 *For any graph G,*

$$\Delta(G) \leq \chi'(G) \leq \Delta(G) + 1.$$

Proof. We already know $\Delta(G) \leq \chi'(G)$. Let us assume that $\chi'(G) > \Delta + 1$. Select a maximal $(\Delta + 1)$-painting π of G and a vertex x such that $s(x) < d(x)$. There must exist colors c_1 and c_0 such that x lies on two edges of color c_1 and no edge of color c_0.

Let xy_1 be an edge of color c_1. Since $d(y_1) < \Delta + 1$, there is some color c_2 which is not represented at y_1; but c_2 must be represented at x, for otherwise we could construct a k-painting of greater order than π by recoloring xy_1 with c_2. Select a vertex y_2 such that xy_2 has color c_2. Continuing this process we obtain sequences $c_1, c_2, \cdots$ of colors and $y_1, y_2, \cdots$ of vertices such that xy_i has color c_i and c_{i+1} does not belong to $S(y_i)$ for any i. Since only k colors are available, the sequence of colors must contain repetitions. Say c_q is the first such repetition: q is the smallest integer such that, for some p less than q, $c_p = c_q$.

We now construct a new k-painting ρ from π by recoloring xy_i in c_{i+1} when $1 \leq i < p$, and another new k-painting σ from ρ by recoloring xy_i in c_{i+1} when $p \leq i \leq q - 1$. Obviously both are maximal. Write H_ρ and H_σ for the graphs $C_{c_0}(\rho) \cup C_{c_p}(\rho)$ and $C_{c_0}(\sigma) \cup C_{c_p}(\sigma)$, respectively. By Lemma 8.2, the components of H_ρ and H_σ which contain x must both be odd cycles. Call them K_ρ and K_σ. K_ρ is

$$x, y_{p-1}, \cdots, y_p, x.$$

So y_p has degree 2 in H_ρ, and consequently it has degree 1 in H_σ. But none of the edges in the path $x, y_{p-1}, \cdots, y_p$ were recolored in the passage from ρ to σ, so y_p must lie in K_σ, and have degree 2 in H_σ — a contradiction.   □

Graphs which satisfy $\chi'(G) = \Delta$ are called *class 1*; those with $\chi'(G) = \Delta + 1$ are called *class 2*. Clearly, for regular graphs, being class 1 is equivalent to having a one-factorization. The class 2 graphs include the odd-order complete graphs and the odd cycles. In general, class 2 graphs appear to be relatively rare, and for that reason we shall investigate some properties of class 2 graphs.

Theorem 8.5 [18] *If G has v vertices and e edges and*

$$e > \left\lfloor \tfrac{v}{2} \right\rfloor \Delta(G) \tag{8.1}$$

then G is class 2.

Proof. Suppose G satisfies (8.1) and is class 1. Select a $\Delta(G)$-coloring of G. Obviously no color class can contain more than $\left\lfloor \tfrac{v}{2} \right\rfloor$ edges, so

$$e = \sum c_i \le \left\lfloor \tfrac{v}{2} \right\rfloor \Delta(G)$$

which is a contradiction.   □

The edge e is called *critical* (with respect to edge-coloring) if

$$\chi'(G - e) < \chi'(G).$$

A *critical graph* is defined to be a connected class 2 graph in which every edge is critical. A critical graph of maximal degree Δ is often called Δ-critical. The importance of this index follows from the obvious fact that every class 2 graph contains a Δ-critical subgraph. In fact we can say more:

Theorem 8.6 *If G is a graph of class 2 then G contains a k-critical subgraph for each k such that $2 \le k \le \Delta(G)$.*

Proof. We prove that any class 2 graph G contains a critical graph G_1 with $\Delta(G_1) = \Delta(G)$. We then prove that G_1 contains a class 2 graph H with $\Delta(H) = k$ for each k satisfying $2 \le k < \Delta(G)$. Then H contains a k-critical graph H_1, and H_1 is of course a subgraph of G. If we simply write Δ we shall mean $\Delta(G)$.

First, suppose G is not critical. Then there will be an edge e_1 such that

$$\chi'(G - e_1) = \chi'(G).$$

Write $G_1 = G - e_1$. Clearly $\Delta(G_1) = \Delta$, and G_1 is class 2. Either G_1 is critical or we can delete another edge, e_2 say, and continue. Eventually the process must stop, because of finiteness, so eventually a Δ-critical subgraph is constructed.

Now consider G_1. Let k be any integer satisfying $2 \leq k < \Delta$. Select any edge uv in G_1, and let π be a Δ-edge-coloring of $G_1 - uv$. (Such a π exists because G_1 is critical.) Clearly $S(u) \cup S(v)$ must equal $\{1, 2, \cdots, \Delta\}$ — if any color were missing we could color uv in that color, and Δ-color G_1. But $|S(u)| < \Delta$, because u had degree at most Δ in G and therefore degree at most $\Delta - 1$ in G_1; so there is some color, i say, not in $S(u)$, and similarly there is some color, j say, not in $S(v)$, and $j \neq i$. Without loss of generality we can assume $i = 1$ and $j = 2$. Then write H for the subgraph

$$H = C_1 \cup C_2 \cup \cdots \cup C_k \cup \{uv\}.$$

Clearly $\chi'(H) \geq k + 1$: if H could be colored in k colors then G_1 could be colored in Δ. But no vertex has degree greater than k in H: if x is not u or v, then it lies on at most one edge in each of $C_1, C_2, \cdots, C_k$, while u lies on at most $k - 1$ of those (none in C_1) plus uv, and similarly for v. So $\Delta(H) \leq k$. It follows from Theorem 8.4 that $\Delta(H) = k$, $\chi'(H) = k + 1$, and H is the required class 2 subgraph of maximum degree k. $\qquad\square$

In the discussion of Δ-critical graphs, the vertices of degree Δ are of special significance. Such vertices are called *major*. A vertex x for which $d(x) < \Delta$ is called *minor*.

Suppose xy is a critical edge in a class 2 graph G, and π is a Δ-coloring of $G - xy$. We define a (π, x, y)-*fan*, or simply *fan*, in G to be a sequence of distinct edges $xy_1, xy_2, \cdots, xy_n$ such that the color on edge xy_i does not appear in $S(y_{i-1})$, where, in particular, y_0 is interpreted as y. We write $F(x)$ for the set of all y_i which appear in any (π, x, y)-fan. In the following proofs it is convenient to write $T(v)$ for the complement of $S(v)$, so $T(v)$ is the set of all colors *not* represented at vertex v, and we define P to be the set of all colors not represented on edges from x to $F(x)$.

Lemma 8.7 *The set P is disjoint from $T(y)$ and from each $T(z)$, $z \in F(x)$.*

Proof. Suppose $i \in T(y)$. There must be an edge xz which is colored i, because otherwise we could extend π to a Δ-coloring of G by setting $\pi(xy) = i$. Now xz is a (one-edge) fan, so $i \notin P$. So P and $T(y)$ are disjoint.

Suppose P and $T(z)$ have a common element i. Let us denote a fan containing z by

$$xy_1, xy_2, \cdots, xy_n,$$

where $y_n = z$, and say $\pi(xy_j) = i_j$. If $i \in T(x)$, then one could carry out the recoloring

$$\pi(xy) = i_1, \pi(xy_1) = i_2, \cdots, \pi(xy_{n-1}) = i_n, \pi(xy_n) = i,$$

with other edges unchanged. This yields a Δ-coloring of G. But if $i \notin T(x)$, there is some edge xw such that $\pi(xw) = i$, and i belongs to no fan. However

$$xy_1, xy_2, \cdots, xy_n, xw$$

is clearly a fan — a contradiction. So P is disjoint from each $T(z)$. $\square$

Lemma 8.8 *Suppose $i \in T(x)$ and $j \in T(z)$, where $z \in F(x) \cup \{y\}$. Then x and y belong to the same component of $C_i \cup C_j$.*

Proof. In the case $z = y$ this is easy, because otherwise we could exchange colors i and j in the component which contained x and then color xy with color j. So we assume $z \in F(x)$.

We call a fan $xy_1, xy_2, \cdots, xy_n$ *deficient* if there exist colors i and j such that $i \in T(x)$, $j \in T(y_n)$ and x and y_n belong to different components of $C_i \cup C_j$. Among all deficient fans, consider one for which n has the minimum value. For the relevant i and j, exchange colors i and j in the component containing x. The result is a Δ-coloring of $G - xy$ in which y_n is still a member of $F(x)$ — the minimality of the fan ensures that it is still a fan in the new coloring — and j belongs to $P \cap T(y_n)$, in contradiction of Lemma 8.7. $\square$

Lemma 8.9 *The sets $T(z)$, where $z \in F(x) \cup \{y\}$, are pairwise disjoint.*

Proof. Suppose $j \in T(v) \cap T(w)$, where v and w are in $F(x) \cup \{y\}$, and select $i \in T(x)$. (Since $|T(x)| = d(x) - 1 < \Delta$, such a color i exists.) Then x, v and w all belong to the same component of $C_i \cup C_j$. But the components of this graph are all paths and cycles. Since x, v and w are all of degree 1 in $C_i \cup C_j$, we must have a path with three endpoints, which is impossible. $\square$

Theorem 8.10 *Suppose xy is a critical edge in a class 2 graph G. Then x is adjacent to at least $\Delta - d(y) + 1$ major vertices other than y.*

Proof. In the preceding notation we have, from the Lemmas, that P, $T(y)$ and the $T(z)$, $z \in F(x)$, are pairwise disjoint. Now P has $\Delta - |F(x)|$ elements, $T(y)$ has $\Delta - d(y) + 1$ (since the edge xy received no color), and each $T(z)$ has $\Delta - d(z)$. Since these are disjoint sets of colors,

$$\Delta \geq |P| + |T(y)| + \sum_{z \in F(x)} |T(z)|$$

$$= 2\Delta - |F(x)| - d(y) + 1 + \sum_{z \in F(x)} (\Delta - d(z))$$

whence

$$\Delta - d(y) + 1 \leq \sum_{z \in F(x)} (1 + d(z) - \Delta).$$

Since none of the terms on the right-hand side can be greater than 1, and all are integral, it follows that at least $\Delta - d(y) + 1$ of them must equal 1. So at least $\Delta - d(y) + 1$ of the vertices z in $F(x)$ satisfy $d(z) = \Delta$, and are major. $\square$

Corollary 8.10.1 *In a critical graph every vertex is adjacent to at least two major vertices.*

Proof. In the notation of the Theorem, either $d(y) < \Delta$ (and $\Delta - d(y) + 1 \geq 2$) or $d(y) = \Delta$ (and x is adjacent to at least one major vertex other than y and also to y). $\square$

If we take x to be a major vertex in the preceding Corollary, we see that every critical graph contains at least three major vertices. If G is any class 2 graph, then considering the $\Delta(G)$-critical subgraph we see:

Corollary 8.10.2 *Every class two graph contains at least three major vertices.* $\square$

Corollary 8.10.3 *A critical graph G has at least $\Delta(G) - \delta(G) + 2$ major vertices.*

Proof. In the Theorem, take x to be a major vertex and y to be a vertex of degree $\delta(G)$. $\square$

Theorem 8.10 and the Corollaries were essentially proven by Vizing [152] and are collectively known as Vizing's Adjacency Lemma.

Corollary 8.10.4 *A Δ-critical graph on v vertices has at least $\frac{2v}{\Delta}$ major vertices.*   □

Proof. Let us count the edges adjacent to the major vertices. There are at least $2v$, by Corollary 8.10.1, but there are exactly Δ per major vertex.   □

Corollary 8.10.5 *Suppose w is a vertex of a graph G which is adjacent to at most one major vertex, and e an edge containing w. Then*

$$\Delta(G - e) = \Delta(G) \Rightarrow \chi'(G - e) = \chi'(G)$$

and

$$\Delta(G - w) = \Delta(G) \Rightarrow \chi'(G - w) = \chi'(G).$$

Proof. G is either class 1 or class 2. We treat the two cases separately. If G is of class 1, then

$$\Delta(G) = \chi'(G) \geq \chi'(G - e) \geq \Delta(G - e) = \Delta(G)$$

and therefore $\chi'(G - e) = \chi'(G)$. Similarly $\chi'(G - w) = \chi'(G)$.

If G is of class 2, then, by Theorem 8.6, G contains a $\Delta(G)$-critical subgraph, H say. By Corollary 8.10.4, w cannot be a vertex of $V(H)$. So H is a subgraph of $G - w$, and

$$1 + \Delta(G) = \chi'(H) \leq \chi'(G - w) \leq \chi'(G) = 1 + \Delta(G),$$

so $\chi'(G - w) = \chi'(G)$. Similarly $\chi'(G - e) = \chi'(G)$.   □

We will later need to use the idea of the core of a graph. If G is a graph of maximal degree Δ, the *core* G_Δ is the subgraph induced by the set of all vertices of degree Δ.

Lemma 8.11 *Suppose the core G_Δ of G satisfies the following description.*

   (i) *$V(G_\Delta) = U \cup V \cup W$, a disjoint union into three parts, where $U = \{u_1, u_2, \cdots, u_p\}$ and $V = \{v_1, v_2, \cdots, v_p\}$.*

   (ii) *Whenever $i \leq j$, there is an edge (u_i, v_j).*

   (iii) *Every other edge joins a member of V to a member of $V \cup W$.*

Then G is class 1.   □

The proof is left as an exercise.

We started this chapter by saying that edge-colorings provide a generalization of one-factorizations to graphs which are not regular. The following Theorem (which is proven directly in [29] but is a special case of a Theorem in [67]), uses edge-colorings for a different generalization: the graphs are regular but the factors are not.

Theorem 8.12 *Suppose G is a graph with km edges, where $k \geq \chi'(G)$. Then G can be factored into k factors, each of which is a matching with m edges.*

Proof. We proceed by induction on k. Clearly the Theorem is true for $k = 1$. Suppose it is true for $k = h - 1$, and suppose G has k_1 edges and $\chi'(G) \leq k_1$. Select a h-edge-coloring π of G. Say C_i is the largest color class under π. Then C_i consists of $|C_i|$ disjoint edges of G, and $|C_i| \geq \frac{|E(G)|}{k} = m$. Delete any m edges of C_i from G. The resulting graph has $(h-1)m$ edges and chromatic number at most $h - 1$, so by hypothesis it can be decomposed into $h - 1$ m-edge matchings. The set of edges which were deleted provides the h-th matching for G, as required. $\qquad\square$

Exercises 8

8.1 Prove that $\chi'(K_n) = n$ if n is odd and $\chi'(K_n) = n - 1$ if n is even.

8.2 Prove that if C is a color-class in an edge-coloring of a graph G then

$$|C| \leq \left\lfloor \frac{|V(G)|}{2} \right\rfloor$$

8.3 Prove that all trees are of class 1.

8.4 Prove that every non-empty regular graph with an odd number of vertices is of class 2.

8.5 Suppose G is a Δ-critical graph and vw is an edge of G. Prove that $d(v) + d(w) \geq \Delta + 2$.

8.6 Show that any critical graph is 2-connected.

8.7 Verify that the graph in Figure 8.1 is 3-critical.

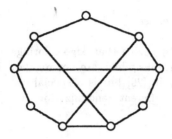

Figure 8.1   A 3-critical graph.

8.8 Suppose G has maximal degree Δ. Prove that $G \backslash G_\Delta$, the graph obtained by deleting the edges of G_Δ from G, can be Δ-colored.

8.9 (*) Prove Lemma 8.11 [39]

9

ONE-FACTORIZATIONS AND TRIPLE SYSTEMS

Among the most important combinatorial designs are *balanced incomplete block designs*. A balanced incomplete block design with parameters (v, b, r, k, λ) is a way of choosing b subsets of size k from a v-set so that any element belongs to precisely r of the sets and any two elements commonly belong to λ of them. The k-sets are called *blocks*. It is easy to see that these parameters are not independent — in fact, $r(k - 1) = \lambda(v - 1)$ and $bk = vr$ — and that the constancy of r is implied by the constancy of k and λ.

We shall discuss only one relatively simple case. A *Steiner triple system* is a balanced incomplete block design with $k = 3$ and $\lambda = 1$. Substituting these values into the given parameter relations we obtain $v = 2r + 1$ and $3b = vr$. Since r and b are clearly integers, v must be congruent to 1 or 3 (mod 6). These conditions are sufficient, as was first shown using a lengthy argument by Kirkman [99].

A Steiner triple system $\mathcal{T}$ is called a *subsystem* of a system $\mathcal{S}$ if every block of $\mathcal{T}$ is a block of $\mathcal{S}$. In other words, $\mathcal{S}$ consists of $\mathcal{T}$ together with some other blocks (each of which will, of course, contain at least one element not in $\mathcal{T}$). A famous result of Doyen and Wilson [61] tells us that a Steiner triple system of order u is a subsystem of some Steiner triple system of order v for every u and v such that u and v are congruent to 1 or 3 modulo 6 and $v > 2u$ (the condition $v > 2u$ is easily seen to be necessary).

In this chapter, we first give a very easy construction to prove the existence of a Steiner triple system of order $6n + 3$ for every n — the construction is essentially due to Bose [26] but was presented there in a very different form. Then we give a construction to prove the existence of Steiner triple systems of all possible orders, which essentially follows [140]. Finally we exhibit the construction of Stern and Lenz [141] which proves the Doyen-Wilson Theorem. All three proofs are applications of one-factorizations. It is clear that, in an exposition of design theory, only the last would be necessary. However, the three proofs are of increasing degrees of complexity (the first is almost trivial, the last uses Vizing's Theorem), and they are three interesting applications of one-factorizations in design theory.

First suppose $v \equiv 3 \pmod 6$. Say $v = 6n + 3$. Select a near-one-factorization of the K_{2n+1} based on points $\{0, 1, \cdots, 2n\}$; denote the factors by $F_0, F_1, \cdots, F_{2n}$, where F_i omits the point i.

We build a Steiner triple system on a set of $6n + 3$ objects labeled

$$\{0^1, 0^2, 0^3, 1^1, 1^2, \cdots, 2n^3\}.$$

First we construct a set of $2n + 1$ blocks

$$\{0^1, 0^2, 0^3\}, \{1^1, 1^2, 1^3\}, \cdots, \{2n^1, 2n^2, 2n^3\}.$$

Then, for each near-one-factor F_j we construct $3n$ blocks: if

$$F_j = \{x_1 y_1, x_2 y_2, \cdots, x_n y_n\},$$

then the $3n$ blocks are

$$
\begin{array}{cccc}
\{x_1^1, y_1^1, j^2\}, & \{x_2^1, y_2^1, j^2\}, & \cdots, & \{x_n^1, y_n^1, j^2\} \\
\{x_1^2, y_1^2, j^3\}, & \{x_2^2, y_2^2, j^3\}, & \cdots, & \{x_n^2, y_n^2, j^3\} \\
\{x_1^3, y_1^3, j^1\}, & \{x_2^3, y_2^3, j^1\}, & \cdots, & \{x_n^3, y_n^3, j^1\}
\end{array}
$$

It is easy to verify that these $3n(2n + 1)$ blocks ($3n$ for each j, $2n + 1$ values of j), together with the $2n + 1$ blocks constructed earlier, form a Steiner triple system. So we have a very easy proof of

Theorem 9.1 *There is a Steiner triple system of order v whenever $v \equiv 3$ (mod 6).*   □

The easiest proof that Steiner triple systems exist for all possible orders combines the above result with an equally simple proof for the case $v \equiv 1 \pmod 6$. However, there is a slightly longer recursive proof that involves one-factorizations throughout. It uses two Lemmas.

Lemma 9.2 *If there is a Steiner triple system* T *with* u *elements, then there is a Steiner triple system* S *with* $2u + 1$ *elements.*

Proof. Suppose T is a triple system with the u elements $\{u+1, u+2, \cdots, 2u\}$. Since u must be odd, there is a one-factorization $\mathcal{F} = \{F_1, F_2, \cdots, F_u\}$ of K_{u+1}, which we take to have vertex-set $\{0, 1, \cdots, u\}$. For each $i, 1 \leq i \leq u$, form the $\frac{u+1}{2}$ triples $\{x, y, u+i\}$, where xy is an edge of F_i. Since $\mathcal{F}$ is a one-factorization, these triples will contain all the pairs $\{x, y\}$ where $0 \leq x \leq u$ and $0 \leq y \leq u$ once each. Since F_i is a one-factor, the pair $\{x, u+i\}$ is contained in exactly one triple for each $x, 0 \leq x \leq u$. If these triples are taken together with the blocks of T, they form the required Steiner triple system. $\square$

Lemma 9.3 *If there is a Steiner triple system* T *with* u *elements, then there is a Steiner triple system* S *with* $2u + 7$ *elements.*

Proof. We first consider the $u+7$ elements $\{0, 1, \cdots, u+6\}$, which we interpret as integers modulo $u+7$. Since $u+7$ is even, we can write $u+7 = 2w$. The $u+7$ triples $\{x, x+1, x+3\}$, where $0 \leq x \leq u+6$, are clearly disjoint, and between them contain all pairs $\{x, y\}$ where $x - y = \pm 1, \pm 2$ or ± 3, once each. If we interpret the elements as vertices of a graph, the edges not contained in any of these triples make up a graph $G(w - 2, w)$, which we proved in Theorem 3.4 to have a one-factorization. Let $F_1, F_2, \cdots, F_u$ be the factors in such a one-factorization of $G(w - 2, w)$. With each of the symbols $u + 8, u + 9, \cdots, 2u + 7$, associate one of the factors — F_1 with $u + 8$, F_2 with $u + 9$, and so on — and form triples: from F_i, form all the triples $\{x, y, u+7+i\}$, where xy is in F_i. Finally, append to all the triples so far constructed the blocks of T, which is taken to have elements $\{u + 8, u + 9, \cdots, 2u + 7\}$. The result is a Steiner triple system. $\square$

Theorem 9.4 *There are Steiner triple systems of orders* $6n + 1$ *and* $6n + 3$ *for every non-negative integer* n.

Proof. We proceed by induction. The Theorem is trivially true for $n = 0$. Suppose it is true for $0 \leq n < k$. We must prove there are Steiner triple systems of orders $6k + 1$ and $6k + 3$.

If k is even, there are systems of orders $\frac{k-2}{2}$ and $\frac{k}{2}$, by hypothesis. Now

$$2\left[6\left(\tfrac{k-2}{2}\right) + 3\right] + 7 = 6k + 1,$$

so from Lemma 9.3 there is a system of order $6k + 1$, and

$$2\left[6(\tfrac{k}{2}) + 1\right] + 1 = 6k + 3.$$

Therefore the system of order $6k + 3$ follows from Lemma 9.2. If k is odd, there is a system of order $\frac{k-1}{2}$; since

$$2\left[6(\tfrac{k-1}{2}) + 3\right] + 1 = 6k + 1,$$
$$2\left[6(\tfrac{k-1}{2}) + 1\right] + 7 = 6k + 3,$$

the result again follows from the Lemmas.   □

The proof of Theorem 9.4 actually shows the existence of a Steiner triple system with a relatively large subsystem for each order. However, rather more is required for the Doyen-Wilson Theorem. We give a proof which follows [141].

A *cyclic* graph of order m is a graph whose vertices are the integers modulo m, with the property that if (x, y) is an edge then so is $(x+i, y+i)$ for $1 \le i \le m-1$. Obviously cyclic graphs are regular. One can specify a cyclic graph by listing all the edges $(0, d)$ contained in it; we shall call the set $\{d : (0, d) \in G\}$ the *basis* of a cyclic graph G. An *even* cyclic graph is a cyclic graph of even order, say $m = 2w$, with w in its basis.

Lemma 9.5 *Every even cyclic graph has a one-factorization.*

Proof. Suppose G is an even cyclic graph. If $\gcd(2w, d)$ is even, then the graph with edges $\{(i, i + d) : 0 \le i \le 2w - 1\}$ consists of even cycles, and has a one-factorization. So it is sufficient to one-factor a graph derived from G by deleting some elements d from the basis for which $\frac{2w}{\gcd(2w,d)}$ is even. We write H for the graph derived from G by deleting from the basis all such elements d except for $d = w$, and H' for the graph derived from H by deleting the edges $(i, i + w)$.

Suppose 2^t divides $2w$ but 2^{t+1} does not. In order that $\frac{2w}{\gcd(2w,d)}$ be odd, 2^t must divide d. It follows that, if x is adjacent to y in H', then 2^t must divide $y - x$. Therefore, if there is a path from x to y in H', 2^t must divide $y - x$. So x and $x + w$ must lie in different components of H', since 2^t does not divide w. Consequently H' can be partitioned into two (not necessarily connected) disjoint subgraphs, E and F say, such that for every x one of $\{x, x + w\}$ is a vertex in E and the other is in F. Moreover, the cyclic property means that (x, y) is an edge of E if and only if $(x + w, y + w)$ is an edge of F.

Suppose H is regular of degree $s + 1$. Then H' is regular of degree s, so $\Delta(E) \leq s$. By Theorem 8.4, E has an $(s + 1)$-edge-coloring. Select such a coloring, and color F by applying to $(x + w, y + w)$ the color which (x, y) received in E. There will be a color, c_x say, which does not appear on any edge touching x (or, consequently, $x + w$). Apply color c_x to edge $(x, x + w)$. The result is a $(t + 1)$-coloring of H. So H (and consequently G) has a one-factorization. $\square$

Suppose S is any set of triples. The *leave* of S is defined to be the graph whose vertices are the elements of the triples, in which two vertices are adjacent if and only if they do not occur together in any triple of S.

Lemma 9.6 *Suppose there is a Steiner triple system of order u and there is a set S of triples with elements in $\mathbb{Z}_{2w}$ whose leave is an even cyclic graph of degree u. Then there is a Steiner triple system of order $u + 2w$ with a subsystem of order u.*

Proof. Relabel the elements on which S is based as $u + 1, u + 2, \cdots, u + 2w$, and denote the one-factors in a one-factorization of its leave as $F_1, F_2, \cdots, F_u$. Then the triples $\{i, x, y\}$, where xy ranges through all edges of F_i and i ranges from 1 to u, together with the triples in a triple system based on $\{1, 2, \cdots, u\}$, constitute a triple system based on $\{1, 2, \cdots, u + 2w\}$. $\square$

As an example of how the conditions of the Lemma can conveniently be satisfied, we define a *cyclic system* of order $2w$ to be a set of triples on $\mathbb{Z}_{2w}$ with the property that if $\{x, y, z\}$ is a triple then so is $\{x + i, y + i, z + i\}$ for every i. A cyclic system with $(0, w)$ in its leave satisfies the conditions. A cyclic system of order $2w$ whose leave is an even cyclic graph of degree u is called a $CS(u, 2w)$.

Theorem 9.7 *There is a Steiner triple system of order v with a subsystem of order u whenever $u \equiv 1$ or $3 \pmod 6$ and $v \equiv 1$ or $3 \pmod 6$ and $v > 2u$.*

Proof. The cases $u = 1$, $u = 3$ and the particular case $u = 9, v = 19$ are covered by the earlier Theorems. For the other cases we need two specific cyclic systems. They are defined in terms of *starter blocks*; if the block $\{0, x, y\}$ is listed then the design contains the block $\{i, x + i, y + i\}$ for $0 \leq i \leq 2w - 1$. S_1 is defined when $2w \geq 12t$; it has starter blocks

$$\{0, 3t - 1, 3t\} \qquad\qquad \{0, 5t - 1, 5t\}$$
$$\{0, 3t - 2, 3t + 1\} \qquad\qquad \{0, 5t - 2, 5t + 1\}$$
$$\cdots$$
$$\{0, 2t, 4t - 1\} \qquad\qquad \{0, 4t + 1, 6t - 1\}$$

and contains all the pairs $\{x, y\}$ where $1 \le y - x \le 6t - 1$ except when $y - x = 4t$ or $5t$. S_2 is defined when $2w \ge 12t + 6$; it has starter blocks

$$\{0, 5t + 2, 5t + 3\} \qquad\qquad \{0, 3t, 3t + 2\}$$
$$\{0, 5t + 1, 5t + 4\} \qquad\qquad \{0, 3t - 1, 3t + 3\}$$
$$\cdots$$
$$\{0, 4t + 3, 6t + 2\} \qquad\qquad \{0, 2t + 1, 4t + 1\}$$

and contains all the pairs $\{x, y\}$ where $1 \le y - x \le 6t + 2$ except when $y - x = 3t + 1$ or $4t + 2$.

(i) Say v and u are not congruent modulo 6. The possibilities are $v - u \equiv 2, 8, 4$ or $10 \pmod{12}$, where $u \equiv 1 \pmod{6}$ in the first two cases and $u \equiv 3 \pmod{6}$ in the others. Write $v - u = 2w$, and form a $CS(u, 2w)$ by deleting an appropriate number of starter blocks from S_1 or S_2. (Use S_1 when $2w \equiv 2$ or $4 \pmod{12}$, S_2 when $2w \equiv 8$ or 10; the "appropriate number" is $\frac{u-7}{6}$ when $u \equiv 1 \pmod{6}$ and $\frac{u-9}{6}$ when $u \equiv 3 \pmod{6}$. It is easy to see that this is possible provided $u \ge 7$ and $v > 2u$.)

(ii) If u and v are congruent modulo 6 this does not work — it would give as the leave a cyclic graph of degree congruent to $5 \pmod 6$. We have to make slightly different modifications in the two cases. We write $v - u = 2w = 6m$; when m is even the modification is carried out on S_1, and when m is odd it is carried out on S_2.

(iia) $v \equiv u \equiv 3 \pmod 6$. Add the triples $\{i, i + 2m, i + 4m\}, 0 \le i \le 2m - 1$, to the design. Then delete any h starter blocks to produce a suitable design for the case $u = 6h + 3$. (In this instance, $u = 3$ is also given by the construction.)

(iib) When $v \equiv u \equiv 1 \pmod 6$, omit the starter block $\{0, m, 2m - 1\}$ and $h - 1$ other starter blocks. Then add the $4m$ blocks

$$\left. \begin{array}{ll} \{i, i + m, i + 2m\} & \{i + 2m, i + 3m, i + 4m\} \\ \{i + m, i + 3m, i + 5m\} & \{i, i + 4m, i + 5m\} \end{array} \right\} 0 \le i \le m - 1$$

The resulting design is not cyclic, but its leave is an even cyclic graph of degree $6h + 1$, so the construction can be used for $u = 6h + 1, h \ge 1$. $\qquad\square$

One can also go from Steiner triple systems to one-factorizations. Suppose a triple system has v elements $x_1, x_2, \cdots, x_v$. Element x_i belongs to $\frac{v-1}{2}$ blocks. Define F_i to consist of the pair ∞i together with all the pairs jk such that $\{x_i, x_j, x_k\}$ is a block. Then $\{F_1, F_2, \cdots, F_v\}$ is a one-factorization of K_{v+1}, and is called a *Steiner factorization*.

One can be quite specific in describing certain Steiner factorizations. If s is any positive integer greater than 2, the finite projective geometry of dimension $s - 1$ over $GF(2)$ forms a Steiner triple system on $2^s - 1$ elements. Those elements can be interpreted as the non-zero binary vectors of length s, and the triples have the form $\{x, y, x + y\}$ where addition is component by component modulo 2. The corresponding one-factorization of K_{2^s} is easily described. The vertices are the 2^s binary vectors of length s. Factor F_x consists of the pairs $\{y, x + y\}$ where y ranges through the possible vectors; in other words yz lies in F_x if and only if $y + z = x$. We shall denote this one-factorization as $\mathcal{GG}_{2^s}$, and call it a *geometric factorization*.

Exercises 9

9.1 Construct Steiner triple systems of orders 9, 13, and 15.

9.2 (A generalization of Lemma 9.5.) Show that if a cyclic graph of order $2w$ contains at least one edge $(0, d)$ where $\frac{2w}{\gcd(d, 2w)}$ is even then it has a one-factorization.

9.3 A *Steiner near-factorization* is a near-one-factorization formed by deleting the element ∞ from a Steiner factorization. Suppose $\mathcal{F}_1$ and $\mathcal{F}_2$ are Steiner near-factorizations constructed from two Steiner triple systems $\mathcal{T}_1$ and $\mathcal{T}_2$ which contain no common triple. Prove that $\mathcal{F}_1$ and $\mathcal{F}_2$ are disjoint (in the sense of Exercise 3.3).

10

STARTERS

The factorization of Theorem 3.2 is easy to visualize in graphical terms. Suppose the vertices of K_{2n} are arranged in a circle with one vertex in the center. This arrangement is shown in Figure 10.1, with the vertices labelled $\{\infty, 0, 1, \cdots, 2n-2\}$. The one-factor F_0 is shown in that Figure. F_1 is constructed by rotating F_0 through one-$(2n-2)$th part of a revolution, clockwise, and subsequent factors are obtained in the same way. It is easy to see that the $n-1$ chords separate pairs of vertices which differ (modulo $2n-1$) by different numbers, and that on rotation the one edge which represents distance d will generate all edges with distance d.

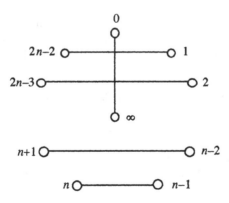

Figure 10.1 The patterned starter.

To generalize this construction we make the following definition. A *starter* in an abelian group G of order $2n - 1$ is an ordered partition of the non-zero members of G into 2-sets $\{x_1, y_1\}, \{x_2, y_2\}, \cdots, \{x_{n-1}, y_{n-1}\}$ with the property that the $2n - 2$ differences $\pm(x_1 - y_1), \pm(x_2 - y_2), \cdots, \pm(x_{n-1} - y_{n-1})$ are all different and therefore contain every non-zero element of G precisely once.

It is clear that the set F_0 used in the proof of Theorem 3.2 is a starter. On the other hand, suppose any other starter F had been used instead of F_0. We construct a set of $2n - 1$ factors by systematically adding elements of G to F (F itself is obtained by adding the zero element). This process is called *developing* F in G; development in the cyclic group of order $2n - 1$ is called *development modulo* $2n - 1$.

For convenience, we order the resulting factors so that the pair $\{\infty, i\}$ occurs in the ith factor. Now consider any unordered pair $\{x, y\}$ where neither x nor y is the symbol ∞. Suppose $x - y = d$. There will be exactly one pair $\{x_i, y_i\}$ in the starter for which $x_i - y_i = \pm d$; for convenience, assume $x_i - y_i = d$. Then

$$x = x_j + (x - x_j), \quad y = y_j + (x - x_j),$$

so (x, y) will occur in the factor formed by adding $x - x_j$ to the starter. So every possible pair occurs at least once in the list of factors. On the other hand, the whole list of factors will contain only $n(2n - 1)$ pairs, n in each factor, and if any one of the $n(2n - 1)$ possible pairs occurred more than once, there would be too many pairs. So each pair occurs precisely once, and we have constructed a one-factorization. We have proven:

Theorem 10.1 *If there is a starter S in a group G of order $2n - 1$, then the $2n - 1$ factors $\{(\infty, x)\} \cup (S + x)$, where x ranges through G, form a one-factorization of K_{2n} called the one-factorization* generated by S. $\qquad\square$

The type of starter we used in Theorem 3.2, consisting of all the pairs $\{i, -i\}$ where i ranges through the non-zero elements of some group, is called a *patterned starter*; $\mathcal{G}K_{2n}$ is the factorization generated by the patterned starter in the cyclic group $\mathbb{Z}_{2n-1}$. The one-factorization of K_{2n} constructed from the patterned starter in G is called the *patterned factorization* on G; when G is not named, it is assumed to be $\mathbb{Z}_{2n-1}$, so "the patterned factorization" often means "$\mathcal{G}K_{2n}$".

Suppose the one-factorization $\mathcal{F}$ is generated from a starter. Of the $2n$ near-one-factorizations which can be constructed from $\mathcal{F}$ by deleting one vertex, we identify the one obtained by deleting ∞ as being derived from the starter. In the case of a patterned starter, this factorization is called the *patterned near-one-factorization* (on the particular group).

If S is a starter in an abelian group G, then the set of pairs $-S$, constructed by replacing each pair $\{x, y\}$ in S by $\{-x, -y\}$, is also a starter. More generally, if the group G has a multiplication operation defined on it — for example, if G is the set of integers modulo $2n - 1$ or if G is the additive group of any integral domain or field — one may define kS by

$$kS = \{(kx, ky) : \{x, y\} \in S\}.$$

The set kS is not always a starter, but in some cases it will be. For example, if S is a starter in the integers modulo $2n - 1$, and if $\gcd(k, 2n - 1) = 1$, then kS is a starter.

For small orders, it is possible to enumerate all starters. For example, let us find all starters in the group $\mathbb{Z}_9$ of integers modulo 9. There must be a pair whose difference is ± 1. This must be one of the pairs

$$\{1, 2\}, \{2, 3\}, \{3, 4\}, \{4, 5\}, \{5, 6\}, \{6, 7\} \text{ or } \{7, 8\}.$$

Any starter including one of the last three pairs mentioned will be the negative of one involving one of the first three, so we only consider the first four cases.

Suppose a starter includes $\{1, 2\}$. Then it must include:

> one of $\{3, 5\}, \{4, 6\}, \{5, 7\}$;
> one of $\{3, 6\}, \{4, 7\}, \{5, 8\}$;
> one of $\{3, 7\}, \{4, 8\}, \{8, 3\}$.

As repeated elements are to be avoided, it is easy to show that there are two possibilities:

$$\{1, 2\}, \{4, 6\}, \{5, 8\}, \{3, 7\} \tag{10.1}$$
$$\{1, 2\}, \{5, 7\}, \{3, 6\}, \{4, 8\} \tag{10.2}$$

Similar calculations show that there is exactly one starter in each of the next three cases:

$$\{2, 3\}, \{6, 8\}, \{4, 7\}, \{1, 5\}; \tag{10.3}$$
$$\{3, 4\}, \{5, 7\}, \{8, 2\}, \{6, 1\}; \tag{10.4}$$
$$\{4, 5\}, \{8, 1\}, \{3, 6\}, \{7, 2\}.. \tag{10.5}$$

The negatives of all of these are also starters, but the negative of (10.5) equals (10.5) again. So there are nine starters in $\mathbb{Z}_9$.

Suppose $S = \{\{x_1, y_1\}, \{x_2, y_2\}, \cdots\}$ and $T = \{\{u_1, v_1\}, \{u_2, v_2\}, \cdots\}$ are two starters in the same group, and suppose the pairs in T have been labeled so that $v_i - u_i = y_i - x_i$ for all i. Then S and T are called *orthogonal* if:

  (i) $u_i \neq x_i$ for any i;
  (ii) $u_i - x_i \neq u_j - x_j$ when $i \neq j$.

The reason for this definition is the following easy theorem:

Theorem 10.2 [90] *If S and T are orthogonal starters then the one-factorizations generated from them are orthogonal. Conversely, if orthogonal one-factorizations are generated from starters in the same group, then those starters are othogonal.*   □

A starter

$$S = \{\{x_1, y_1\}, \{x_2, y_2\}, \cdots\}$$

is called *strong* if the sums $x_i + y_i$ are all different and non-zero.

Theorem 10.3 *If S is a strong starter in a group G, and T is the patterned starter in G, then S and T are orthogonal.*

Proof. Write $S = \{\{x_1, y_1\}, \{x_2, y_2\}, \cdots\}$. Since G has an odd number of elements, $\{h + h : h \in G\} = G$, so to every element g of G there corresponds an element h such that $h + h = g$; call this element $\frac{1}{2}g$. Define $h_i = \frac{1}{2}(y_i - x_i)$ for each i. Since S is a starter, the elements $\pm(y_i - x_i)$ are all the non-zero elements of G, so $T = \{\{h_1, -h_1\}, \{h_2, -h_2\}, \cdots\}$. Now $x_i - h_i = -\frac{1}{2}(x_i - y_i)$, so $x_i = h_i$ would imply $x_i - y_i = 0$, which cannot be true in a starter, and moreover the $x_i - y_i$ are all different in the starter S, so the $x_i - h_i$ are all different. Therefore S and T are orthogonal.   □

Theorem 10.4 [90] *If S is a strong starter, then $-S$ is a strong starter, and S and $-S$ are orthogonal.*

Proof. It is easy to see that $-S$ is a starter, and the fact that the $x_i + y_i$ are all different implies that the sums from $-S$, $-x_i - y_i$, are all different, so $-S$ is strong.

Again we write $S = \{\{x_i, y_i\}\}$. When T is replaced by $-S$, the conditions (i) and (ii) for orthogonality become:

(i) $-x_i \neq x_i$ for all i;

(ii) $-x_i - x_i \neq -x_j - x_j$ when $i \neq j$.

The first condition is equivalent to $x_i \neq 0$, and the second is equivalent to $x_i \neq x_j$. Both are true because S is a starter. $\quad\square$

Corollary 10.4.1 *If there is a strong starter in a group of order $2n - 1$, then there is a set of three pairwise orthogonal starters in that group, and there is a $3 - POSLS(2n - 1)$.* $\quad\square$

Strong starters were developed in the study of Room squares, and their usefulness there is immediately apparent: it is very easy to check whether a starter is strong, and a strong starter generates a Room square. One important construction for strong starters is due to Mullin and Nemeth [119]. Suppose q is a prime power congruent to $3 \pmod 4$; write R for the set of non-zero quadratic elements in $GF(q)$ and N for the set of non-quadratic elements. Define S_a to consist of all the pairs $\{r, ar\}$ where r ranges through R.

Theorem 10.5 *The set S_a is a starter in the additive group $GF(q)$ whenever a belongs to N. If $a = -1$, S_a is the patterned starter; otherwise it is a strong starter.*

Proof. To show that S_a is a starter, we must prove that it contains all non-zero elements of $GF(q)$ and that the $2(q-1)$ differences generated are distinct. But when r ranges through R, ar ranges through N, for any non-quadratic element a. The differences are $\pm(1 - a)r$, where r ranges through R. Since $q \equiv 3 \pmod 4$, -1 is not quadratic, so $\{-r : r \in R\} = N$. Consequently

$$\{\pm(1 - a)r : r \in R\} = \{(1 - a)r : r \in GF(q)\backslash\{0\}\}.$$

Since 1 is quadratic, $1 - a$ is non-zero and has an inverse, so the set is just $GF(q)\backslash\{0\}$.

When $a = -1$, the starter is clearly the patterned starter. Otherwise $1 + a$ is non-zero, and has an inverse. So $r + ar = (1 + a)r$ is never zero, and $r_1 + ar_1 = r_2 + ar_2$ would imply that $r_1 = r_2$. Therefore different members of S_a have different sums, and S_a is strong. $\quad\square$

The starters S_a are called *Mullin-Nemeth* starters.

Theorem 10.6 *If S_a and S_b are Mullin-Nemeth starters, and $a \neq b$, then S_a and S_b are orthogonal.*

Proof. Suppose $S_a = \{\{x_1, ax_1\}, \{x_2, ax_2\}, \cdots\}$ and $S_b = \{\{u_1, bu_1\}, \{u_2, bu_2\}, \cdots\}$ where $x_i - ax_i = u_i - bu_i$. We need to show that $u_i \neq x_i$ for all i and $u_i - x_i \neq u_j - x_j$ when $i \neq j$. Now

$$u_i \doteq x_i(1-a)(1-b)^{-1};$$

since a and b are different, $(1-b)^{-1} \neq (1-a)^{-1}$, so u_i and x_i are always different. This same fact, that $(1-a)(1-b)^{-1} \neq 1$, tells us that when $x_i \neq x_j$,

$$x_i[(1-a)(1-b)^{-1} - 1] \neq x_j[(1-a)(1-b)^{-1} - 1],$$

that is,

$$u_i - x_i \neq u_j - x_j. \qquad \square$$

Corollary 10.6.1 *If q is a prime power congruent to $3 \pmod 4$ then $\nu(q) \geq \frac{q-1}{2}$.* $\qquad \square$

The Mullin-Nemeth starters were generalized by Dinitz [47], in order to obtain better lower bounds on the number of pairwise orthogonal Latin squares of a given side. In particular his results improve the bounds when the order is congruent to 1 (mod 4).

It is clear that no group of order 3 or 5 can contain a strong starter. Various other groups do not admit of a strong starter — for example, $\mathbb{Z}_3 \times \mathbb{Z}_3$ does not have one (see Exercise 10.6). On the other hand, there are several constructions for strong starters. We give one recursive example due to Horton [89]; another construction is given by Gross [79].

Theorem 10.7 *If G is a finite abelian group of order prime to 6, which admits of a strong starter, then there is a strong starter in the direct product of G with the cyclic group of order 5.*

Proof. Let us write $t = 2n + 1$ for the order of G. Since G is finite and abelian, it is a direct sum of cyclic groups; we can interpret any cyclic group of order m as the ring of integers modulo m under addition, and consider G as the additive group of the direct product of these rings. In this way we endow G with a multiplication. This multiplication will have an identity element, 1 say; we write $1 + 1$ as 2 and $1 + 1 + 1$ as 3, and 2^{-1} and 3^{-1} will exist since $\gcd(t, 6) = 1$.

Let the strong starter in G be

$$S = \left\{ \{x_1, y_1\}, \{x_2, y_2\}, \cdots, \{x_n, y_n\} \right\}.$$

Choose two non-zero elements a and b of G such that neither a nor b equals $x_i + y_i$ for any i. (This requires that $n + 2 \le t - 1$; as no strong starter exists in a group of size less than 7, there is no problem.) Write $h = 2^{-2}(b - a)$ and $g = 2^{-1}a$. Finally, partition the set of all non-zero elements of G into two classes P and N, in such a way that

$$
\begin{aligned}
h &\in P \\
-3^{-1}h &\in P \\
x \in P &\iff -x \in N.
\end{aligned}
$$

We now write down six sets, A, B, C, D, E, F, of unordered pairs of elements of $G \times \mathbb{Z}_5$, where the cyclic group $\mathbb{Z}_5$ of order 5 is written as

$$\mathbb{Z}_5 = \{0, 1, 2, 3, 4\}.$$

(We are using 1, 2 and 3 in two senses, but elements of $\mathbb{Z}_5$ occur only in the right-hand side of ordered pairs, so no confusion arises.) The union of these sets will be shown to be a strong starter. In each description, g ranges through G.

$$
\begin{aligned}
A &= \left\{ \{(x, 0), (y, 0)\} : \{x, y\} \in X \right\} \\
B &= \left\{ \{(x + g, 1), (2x + g, 2)\} : x \in P, x \ne h \right\} \\
C &= \left\{ \{(x + g, 4), (2x + g, 3)\} : x \in P, x \ne h \right\} \\
D &= \left\{ \{(x + g, 1), (2x + g, 3)\} : x \in N \right\} \\
E &= \left\{ \{(x + g, 4), (2x + g, 2)\} : x \in N \right\} \\
F &= \left\{ \{(h + g, 1), (g, 2)\}; \{(h + g, 4), (g, 3)\}; \right. \\
&\qquad \left. \{(g, 1), (g, 4)\}; \{(2h + g, 2), (2h + g, 3)\} \right\}.
\end{aligned}
$$

To prove that this is a starter, we must show that every non-zero element of $G \times \mathbb{Z}_5$ occurs in one pair and that the set of all differences between members of a pair also consists of all non-zero elements.

Every element of the form $(y, 0)$, with $x \neq 0$, occurs in a pair in A. The pairs of B and D will contain all elements of the form $(x + g, 1)$ except for $(h + g, 1)$ and $(0 + g, 1)$; the latter elements arise in F. If x runs through G then so does $x + g$, so we have every element of the form $(y, 1)$. Similarly, all elements $(y, 2)$, $(y, 3)$ and $(y, 4)$ are included.

Since X is a starter, the elements $x - y$, where $\{x, y\} \in X$, comprise $G \setminus \{0\}$. So all elements of the form $(x, 0), x \neq 0$, come up as differences in A. The differences in B contain all the elements of the form $(x, 1)$ where $x \in P$ but $x \neq h$. From C we get all the $(-x, 1)$ with x in the same range; these will equal the $(x, 1)$ with $x \in N$ but $x \neq -h$, since N consists of the negatives of elements of P. The remaining elements $(h, 1)$, $(-h, 1)$ and $(0, 1)$ come from F. Similarly, all elements with second component 2, 3 or 4 arise as differences.

In both cases, to see that 0 does not come up and that no element appears twice, it is sufficient to observe that there are only $5n + 2$ pairs in the union, and consequently $5t - 1$ elements in each of the sets we have been discussing.

We complete the proof by showing that the starter is strong, which means that the $5n + 2$ elements formed by summing the two members of the pair are all distinct and non-zero. The sums from A have the form $(x + y, 0)$ where $\{x, y\} \in X$; since X is strong, these are all different and non-zero. From B and C we get all pairs of the form

$$(3x + 2g, 3), (3x + 2g, 2),$$

where x ranges through $P \setminus \{h\}$. Sets D and E supply all the elements

$$(3x + 2g, 4), (3x + 2g, 1),$$

for $x \in N$. These are all non-zero, and are distinct unless $3x + 2g = 3x_1 + 2g$ where $x \neq x_1$. But 3^{-1} exists, so

$$3x + 2g = 3x_1 + 2g \Rightarrow 3x = 3x_1 \Rightarrow x = x_1.$$

Finally we have the pairs from F. These have sums $(h + 2g, 3)$, $(h + 2g, 2)$, $(2g, 0)$ and $(4h + 2g, 0)$. If the first two have already arisen then

$$h + 2g = 3x + 2g$$

for some x in $P \setminus \{h\}$. But this would imply $3^{-1}h \in P$, and we have specified $3^{-1}h \in N$. We can rewrite $(2g, 0)$ as $(a, 0)$ and $(4h + 2g, 0)$ as $(b, 0)$, and these are not yet in our list since neither a nor b is a sum of a pair in X.

Therefore the union is a strong starter. □

Exercises 10

10.1 Prove the existence of exactly one starter in the cyclic group of order 3, exactly one starter in the cyclic group of order 5, and exactly three starters in the cyclic group of order 7.

10.2 Suppose S is a starter in $\mathbb{Z}_{2n-1}$.

    (i) Prove that kS is a starter if and only if $\gcd(k, 2n-1) = 1$.

    (ii) Starters S and T are called *equivalent* if $T = kS$ for some S. Prove that the patterned starter is not equivalent to any other starter.

    (iii) Prove that there are precisely two equivalence classes of starters in $\mathbb{Z}_7$, and three equivalence classes in $\mathbb{Z}_9$.

10.3 Find all starters in $\mathbb{Z}_3 \times \mathbb{Z}_3$.

10.4 Prove Theorem 10.2.

10.5 Find a strong starter in $\mathbb{Z}_9$.

10.6 Prove that there is no strong starter in $\mathbb{Z}_3 \times \mathbb{Z}_3$.

11

INVARIANTS OF
ONE-FACTORIZATIONS

It is obvious that there is only one one-factor in K_2, and it (trivially) constitutes the unique one-factorization. Similarly K_4 has just three one-factors, and together they form a factorization.

The situation concerning K_6 is more interesting. There are fifteen one-factors of K_6, namely

$$
\begin{array}{lll}
01\ 23\ 45, & 01\ 24\ 35, & 01\ 25\ 34 \\
02\ 13\ 45, & 02\ 14\ 35, & 02\ 15\ 34 \\
03\ 12\ 45, & 03\ 14\ 25, & 03\ 15\ 24 \\
04\ 12\ 35, & 04\ 13\ 25, & 04\ 15\ 23 \\
05\ 12\ 34, & 05\ 13\ 24, & 05\ 14\ 23
\end{array}
$$

(brackets and commas are omitted for ease of reading).

Consider the factor

$$F_1 = 01\ 23\ 45.$$

We enumerate the one-factorizations which include F_1. The only factors in the list which contain 02 and which have no edge in common with F_1 are

$$
\begin{array}{rcl}
F_{21} & = & 02\ 14\ 35, \\
F_{22} & = & 02\ 15\ 34.
\end{array}
$$

We now construct all one-factorizations which contain F_1 and F_{21}; those which contain F_1 and F_{22} can be obtained from them by exchanging the symbols 4 and 5 throughout. But one easily sees that there is exactly one way of completing $\{F_1, F_{21}\}$ to a one-factorization, namely

$$
\begin{array}{ccc}
03 & 15 & 24 \\
04 & 13 & 25 \\
05 & 12 & 34
\end{array}
$$

So there are exactly two one-factorizations containing F_1. It follows (by permuting the symbols $\{0, 1, 2, 3, 4, 5\}$) that there are exactly two one-factorizations containing any given one-factor.

Now let us count all the ordered pairs $(F, \mathcal{F})$ where F is a one-factor of K_6 and $\mathcal{F}$ is a one-factorization containing F. From what we have just said the total must be $2 \cdot 15$, that is, 30. On the other hand, suppose there are N one-factorizations. Since there are 5 factors in each, we have $5N = 30$, so $N = 6$; there are six one-factorizations of K_6.

The above calculations are simple in theory, but in practice they soon become unwieldy. There are 6240 one-factorizations of K_8 (see Exercise 11.1), 1,255,566, 720 of K_{10} ([73]) and 252,282,619,805,368,320 of K_{12} ([51]); no larger numbers are known. In any case, the importance of such numbers is dubious. It is more interesting to know about the existence of "essentially different" one-factorizations. As usual, the first step is to make a sensible definition of isomorphism of factorizations.

Suppose G is a graph which has a one-factorization. Two one-factorizations $\mathcal{F}$ and $\mathcal{H}$ of G, say

$$
\begin{aligned}
\mathcal{F} &= \{F_1, F_2, \cdots, F_k\}, \\
\mathcal{H} &= \{H_1, H_2, \cdots, H_k\},
\end{aligned}
$$

are called *isomorphic* if there exists a map ϕ from the vertex-set of G onto itself such that

$$
\{F_1\phi, F_2\phi, \cdots, F_k\phi\} = \{H_1, H_2, \cdots, H_k\},
$$

where $F_i\phi$ is the set of all the edges $\{x\phi, y\phi\}$ such that $\{x, y\}$ is an edge in $\mathcal{F}$. In other words, when ϕ operates on one of the factors in $\mathcal{F}$, it produces one of the factors in $\mathcal{H}$. The map ϕ is called an *isomorphism*. In the case where $\mathcal{H}$ equals $\mathcal{F}$ itself, ϕ is called an *automorphism*; the automorphisms of a one-factorization $\mathcal{F}$ form a group, the *automorphism group* of $\mathcal{F}$, which is written $Aut(\mathcal{F})$. The set of all isomorphisms from $\mathcal{F}$ to $\mathcal{H}$ will be denoted $IG(\mathcal{F}, H)$.

Now that isomorphism is defined, we can more formally state our question about "essentially different" factorizations: what is the number of isomorphism classes (equivalence classes under isomorphism) of one-factorizations of K_{2n}?

This is a very difficult question, and exact results are known only for $2n \leq 12$. There is a unique one-factorization of K_{2n}, up to isomorphism, for $2n = 2$, 4 or 6. There are exactly six for K_8; these were found by Safford (see [46]) and a full exposition is given in [167]. The six factorizations are shown in Table 11.1. For K_{10} the number of one-factorizations up to isomorphism is 396; they were listed in [73] (see also [74]) and are given in Appendix A below. There are 526,915,620 isomorphism classes of one-factorizations of K_{12}; see [51].

To discuss isomorphism of factorizations, it is convenient to use some *invariants* — properties which are the same for all members of an isomorphism class. We shall discuss four invariants: divisions, cycle profiles, trains and tricolor vectors.

First, we introduce the idea of a *division*: If F_1 and F_2 are two factors in a one-factorization of K_{2n} then their union may be a Hamilton cycle; otherwise it must decompose into two or more disjoint parts, and we say F_1 and F_2 constitute a *division*, or 2-division, in the one-factorization. More generally a *k-division* is a set of k one-factors whose union is disconnected. It is useful to observe that each of the connected parts (or *components*) in a k-division must have at least $k + 1$ vertices.

It is clear that the image of a d-division, under any permutation of the vertices, remains a d-division. So "having a d-division" is invariant under isomorphism. We shall say a d-division is *maximal* if it cannot be embedded in a $(d + 1)$-division, and write α_d or $\alpha_d(\mathcal{F})$ for the number of maximal d-divisions in a given one-factorization $\mathcal{F}$; then the α_i form a convenient set of invariants for use in discussion of one-factorizations.

We now use the cycle structure to prove the existence of non-isomorphic one-factorizations of K_{2n} when n is an integer greater than 3. It is convenient to work in the integers modulo $2n - 1$. If i and k are any integers we write d_{ik} for the greatest common divisor of $i - k$ and $2n - 1$, that is

$$d_{ik} = \gcd(i - k, 2n - 1),$$

and we define ν_{ik} by

$$d_{ik} \cdot \nu_{ik} = 2n - 1.$$

If $\mathcal{P}_{2n} = \{P_0, P_1, \cdots, P_{2n-2}\}$ is the patterned one-factorization $\mathcal{G}K_{2n}$ in $\mathbb{Z}_{2n-1}$, defined as usual by

$$P_i = (\infty, i), (i - 1, i + 1), (i - 2, i + 2), \cdots, (i - n + 1, i + n - 1),$$

then it is easy to prove the following Lemma:

$\mathcal{F}_1$
$A = 01\ 23\ 45\ 67$
$B = 02\ 13\ 46\ 57$
$C = 03\ 12\ 47\ 56$
$D = 04\ 15\ 26\ 37$
$E = 05\ 14\ 27\ 36$
$F = 06\ 17\ 24\ 35$
$G = 07\ 16\ 25\ 34$

$\mathcal{F}_2$
$A = 01\ 23\ 45\ 67$
$B = 02\ 13\ 46\ 57$
$C = 03\ 12\ 47\ 56$
$D = 04\ 15\ 26\ 37$
$E = 05\ 14\ 27\ 36$
$F = 06\ 17\ 25\ 34$
$G = 07\ 16\ 24\ 35$

$\mathcal{F}_3$
$A = 01\ 23\ 45\ 67$
$B = 02\ 13\ 46\ 57$
$C = 03\ 12\ 47\ 56$
$D = 04\ 16\ 25\ 37$
$E = 05\ 17\ 26\ 34$
$F = 06\ 14\ 27\ 35$
$G = 07\ 15\ 24\ 36$

$\mathcal{F}_4$
$A = 01\ 23\ 45\ 67$
$B = 02\ 13\ 46\ 57$
$C = 03\ 12\ 47\ 56$
$D = 04\ 16\ 27\ 35$
$E = 05\ 17\ 26\ 34$
$F = 06\ 14\ 25\ 37$
$G = 07\ 15\ 24\ 36$

$\mathcal{F}_5$
$A = 01\ 23\ 45\ 67$
$B = 02\ 13\ 47\ 56$
$C = 03\ 14\ 27\ 56$
$D = 04\ 16\ 25\ 37$
$E = 05\ 17\ 26\ 34$
$F = 06\ 12\ 35\ 47$
$G = 07\ 15\ 24\ 36$

$\mathcal{F}_6$
$A = 01\ 23\ 45\ 67$
$B = 02\ 14\ 36\ 57$
$C = 03\ 16\ 25\ 47$
$D = 04\ 17\ 26\ 35$
$E = 05\ 12\ 37\ 46$
$F = 06\ 15\ 27\ 34$
$G = 07\ 13\ 24\ 56$

Table 11.1   One-factorizations of K_8.

Lemma 11.1 *If P_i and P_k are different factors in $\mathcal{P}_{2n}$ then $P_i \cup P_k$ consists of a cycle of length ν_{ik+1} and $\frac{1}{2}(d_{ik} - 1)$ cycles each of length $2\nu_{ik}$.* □

As an immediate consequence of this Lemma we have the following result, which will be used in Chapters 12 and 16.

Corollary 11.1.1 *The union of the two factors P_i and P_j is a Hamilton cycle in K_{2n} if and only if $i - k$ and $2n - 1$ are coprime. In particular $P_i \cup P_{i+1}$ and $P_i \cup P_{i+2}$ are always Hamilton cycles. If $2n - 1$ is prime, any two of the factors form a Hamilton cycle.* □

Theorem 11.2 *When $n > 2$, $\mathcal{P}_{2n}$ cannot contain an $(n - 1)$-division.*

Proof. An $(n - 1)$-division would have two components of order n. Suppose $n > 2$, so that $n - 1 \geq 2$, and let P_i and P_k be two different factors in an $(n - 1)$-division. The 2-division $\{P_i, P_k\}$ has one component of size ν_{ik+1} and $\frac{1}{2}(d_{ik} - 1)$ components of size $2\nu_{ik}$. So one of the components of the $(n - 1)$-division must be a union of disjoint sets of size ν_{ik}. So ν_{ik} divides n; since ν_{ik} also divides $2n - 1$ we have $\nu_{ik} = 1$ and $d_{ik} = 2n - 1$, which is impossible since $1 \leq i, k \leq 2n - 1$ and $i \neq k$. □

Theorem 11.3 *If $n \neq 5$ then no 2-division of $\mathcal{P}_{2n}$ has a component of order $2n - 4$.*

Proof. Consider the 2-division $\{P_i, P_k\}$. Its component cycles have sizes $2\nu_{ik}$ and $\nu_{ik} + 1$. Since ν_{ik} divides the odd number $2n - 1$, ν_{ik} must be odd; and since ν_{ik} cannot equal 1, $\nu_{ik} \geq 3$. If $\nu_{ik} + 1 = 2n - 4$ we have $\nu_{ik} = 2n - 5$, so ν_{ik} divides $(2n - 1) - (2n - 5)$ and ν_{ik} divides 4, which is a contradiction. So the component of order $\nu_{ik} + 1$ is not a component of order $2n - 4$. If the 2-division is to contain a component of order $2n - 4$, then $2\nu_{ik} = 2n - 4$, and ν_{ik} divides $\gcd(2n - 4, 2n - 1)$. So ν_{ik} divides 3, and necessarily $\nu_{ik} = 3$, whence $2n - 4 = 6$ and $n = 5$. Thus $n \neq 5$ implies that the 2-division has no component cycle of order $2n - 4$. □

Theorem 11.4 [154] *For all $n \geq 4$, there exist non-isomorphic one-factorizations of K_{2n}.*

Proof. When $n = 4$, the result follows from the existence of six non-isomorphic factorizations. To complete the proof, we exhibit:

(i) a one-factorization $\mathcal{H}_{2n}$ of K_{2n} which contains an $(n-1)$-division, for every even n;

(ii) a one-factorization $\mathcal{L}_{2n}$ of K_{2n} which contains a 2-division with a component of order $2n - 4$, for every odd n greater than 5;

(iii) a one-factorization of K_{10} not isomorphic to $\mathcal{P}_{10}$.

From Theorem 11.2, $\mathcal{H}_{2n}$ is not isomorphic to $\mathcal{P}_{2n}$; from Theorem 11.3, $\mathcal{L}_{2n}$ is not isomorphic to $\mathcal{P}_{2n}$.

(i) Take $\mathcal{H}_{2n}$ to be the one-factorization $\mathcal{GA}_{2n}$, written in the same notation as Chapter 3. Then the factors

$$\{H_1, H_2, \cdots, H_{n-1}\}$$

form an n-division.

(ii) When n is odd, the factorization $\mathcal{GA}_{2n}$ does not have the property we require, but is easily transformed into one which does. We perform the permutation ϕ,

$$
\begin{aligned}
(n + 1 - j)_\alpha &\mapsto (2j)_\alpha \\
(j + 1)_\alpha &\mapsto (2j + 1)_\alpha \\
1_\alpha &\mapsto 1_\alpha
\end{aligned}
$$

for $j = 1, 2, \cdots, \frac{1}{2}(n-1)$ and $\alpha = 1, 2$, on the J_i of Chapter 3; then for $1 \leq i \leq n$, write $L_i = J_i\phi$. One readily checks that the factors $L_1, L_2, \cdots, L_n$ contain the same edges as $J_1, J_2, \cdots, J_n$. Now

$$
\begin{aligned}
L_1 &= (1_1, 1_2), (2_1, 3_1), (4_1, 5_1), \cdots, ((n-1)_1, n_1), \\
&\quad (2_2, 3_2), \cdots, ((n-1)_2, n_2), \\
L_{n+3} &= (1_1, (n-2)_2), (2_1, (n-1)_2), (3_1, n_2), \\
&\quad (4_1, 1_2), \cdots, (n_1, (n-3)_2).
\end{aligned}
$$

So $L_1 \cup L_{n+3}$ contains the $(2n - 4)$-cycle

$$1_1, 1_2, 4_1, 5_1, 2_2, 3_2, 6_1, 7_1, \cdots, (n-2)_2, 1_1.$$

Therefore $\mathcal{L}_{2n} = \{L_1, L_2, \cdots, L_{2n-1}\}$ is the required factorization.

(iii) Since $\mathcal{P}_{10}$ contains the 3-division $\{F_1, F_4, F_7\}$, it suffices to observe that

$$01\ 23\ 45\ 67\ 89; \quad 04\ 13\ 26\ 58\ 79; \quad 07\ 12\ 34\ 59\ 68$$
$$02\ 14\ 39\ 56\ 78; \quad 05\ 19\ 27\ 38\ 46; \quad 08\ 17\ 25\ 36\ 49$$
$$03\ 18\ 24\ 57\ 69; \quad 06\ 15\ 29\ 37\ 48; \quad 09\ 16\ 28\ 35\ 47$$

contains no 3-division. □

Divisions are of limited usefulness in discussing isomorphism of factorizations, because the number of non-isomorphic one-factorizations of K_{2n} increases more quickly than the number of possible division-structures as n increases. Gelling [73] used a different invariant. Given a one-factorization $\{F_1, F_2, \cdots, F_{2n-1}\}$, write $c_{x,i,j}$ for the length of the cycle through the vertex x in the 2-regular graph $F_i \cup F_j$, and define $c_k(x)$ to equal the number of choices $\{i, j\}$ for which $c_{x,i,j} = k$. In his classification of one-factorizations of K_{10}, Gelling lists the value of $c_4(x)$ for each vertex, and this set of ten numbers, the *cycle profile* distinguishes between the factorizations quite well.

However, the cycle profile and division structure are quite useless in some cases. For example, suppose $F_i \cup F_j$ is a Hamilton cycle in every case — such factorizations are called *perfect*, and will be discussed in Chapter 16. Cycles and divisions will not distinguish between non-isomorphic perfect factorizations. We introduce a third invariant, the train of a one-factorization, which was defined by Dinitz [49] as a modification of an idea introduced for triple systems by White [174].

Suppose $\mathcal{F}$ is a one-factorization of K_{2n}. The *train* $T(\mathcal{F})$ of $\mathcal{F}$ is a directed graph whose vertices are the $n(2n-1)^2$ triples $\{x, y, F\}$, where x and y are (an unordered pair of) vertices of K_{2n} and F is a factor in $\mathcal{F}$. There is exactly one edge leaving each vertex; the edge from $\{x, y, F\}$ goes to $\{z, t, G\}$, where:

edge xz belongs to F;

edge yt belongs to F;

edge xy belongs to G.

(Observe that we allow loops; $\{x, y, F\}$ will have a loop if and only if xy is an edge of F.) It is obvious that isomorphic one-factorizations have isomorphic trains.

We usually simplify trains by considering only the indegree sequence of the train. That is, with a one-factorization $\mathcal{F}$ we associate the sequence $(t_0, t_1, \cdots)$ where t_i equals the number of vertices in the $T(\mathcal{F})$ which have i edges entering. The sequence is written to terminate with the last non-zero element.

One might think that the indegree sequence of a train might be extremely long. However we can prove that it never exceeds $2n$ members.

Theorem 11.5 [59] *The train of any one-factorization of K_{2n} has maximum indegree at most $2n - 1$.*

Proof. Suppose there is an edge from $\{x, y, F\}$ to $\{z, t, G\}$ in $T(\mathcal{F})$. Then F contains either edge xz or yz. If there is also an edge from $\{u, v, F\}$ to $\{z, t, G\}$ then F also contains either edge uz or vz. But there can be only one edge containing z in F. So one of $\{x, y\}$ equals one of $\{u, v\}$. Since xy and uv are edges in the one-factor G, $xy = uv$. Therefore, for a given vertex $\{z, t, G\}$ of $T(\mathcal{F})$, and for a given factor F, there can be at most one edge of the form $\{x, y, F\} \rightarrow \{z, t, G\}$, and the maximum number of edges equals the number of factors in $\mathcal{F}$, $2n - 1$.   □

The maximum indegree of a train is called its *length*. So the above Theorem says that no one-factorization of K_{2n} can have train of length greater than $2n - 1$. In the case of a perfect one-factorization we can say more.

Theorem 11.6 [59] *The train of a perfect one-factorization of K_{2n} has length at most n, for $n > 2$.*

Proof. Suppose $\mathcal{F}$ is a perfect one-factorization of K_{2n}. We prove that, given a vertex $\{z, t, G\}$ of $T(\mathcal{F})$, and given an edge xy of G, there is at most one factor F such that $\{x, y, F\} \rightarrow \{z, t, G\}$. Then the indegree of $\{z, t, G\}$ cannot exceed the number of edges in G, namely n.

Suppose $\{x, y, {}_1 F\} \rightarrow \{z, t, G\}$ and $\{x, y, F_2\} \rightarrow \{z, t, G\}$ in $T(\mathcal{F})$. Either xz and yt or xt and yz are edges of F_1, and either xz and yt or xt and yz are edges of F_2. Since there can be no common edge in F_1 and F_2, their union contains a 4-cycle. This is impossible if $2n > 4$, since $\mathcal{F}$ is perfect.   □

The maximum length $2n - 1$ can be attained for nearly all orders. To prove this we use a special array called an *intercalate square* of side $2n - 2$. This is a symmetric Latin square of side $2n - 2$ whose entries in rows and columns $\{2i - 1, 2i\}$ form a 2×2 Latin square based on $\{2i - 1, 2i\}$. Such an array exists for all even orders except $2n - 2 = 4$; this is easy to see when n is odd (see Exercise 9) and was proven in [71] for the remaining cases.

Theorem 11.7 *There is a one-factorization of K_{2n} whose train has length $2n_1$ whenever $2n \neq 6$.*

Proof. Suppose A is an intercalate square of side $2n - 2$. Define

$$F_0 = \{\infty 01234 \cdots (2n - 3)(2n - 2)\}$$

and when $i > 0$

$$
\begin{aligned}
F_{2n-1} &= \{\infty(2i - 1), 0(2i)\} \cup \{xy : a_{xy} = 2i - 1\}, \\
F_{2n} &= \{\infty(2i), 0(2i - 1)\} \cup \{xy : a_{xy} = 2i\}.
\end{aligned}
$$

Then $\{\mathcal{F}_0, F_1, \cdots, F_{2n-2}\}$ is a one-factorization of K_{2n} and $\{\infty, 0, F_0\}$ has indegree $2n - 1$ in the train.   □

The (unique) one-factorization of K_6 has train with indegree sequence (0,75), of length 1, so the situation is completely determined. One-factorizations with train of maximum length are further discussed in [164].

Finally, another invariant, the *tricolor vector*, was introduced in [77]. Suppose $\mathcal{F} = \{F_0, F_1, \cdots, F_{2n-2}\}$ is a one-factorization of K_{2n}. If a, b and c are three distinct vertices of K_{2n}, the three edges ab, ac and bc must belong to three different factors in $\mathcal{F}$, say F_i, F_j and F_k. Write

$$f(a, b, c) = \{i, j, k\},$$

and define $N(\{i, j, k\})$ to be the number of unordered triples $\{a, b, c\}$ of vertices such that $f(a, b, c) = \{i, j, k\}$. Then the *tricolor vector* of $\mathcal{F}$ is $(v_0, v_1, \cdots)$ where v_n is the number of triples $\{i, j, k\}$ such that $N(\{i, j, k\}) = n$. The first element v_0 is called the *tricolor number* of $\mathcal{F}$.

How well do these invariants distinguish between non-isomorphic factorizations? Trains behave very well. It is shown in [59] that the 396 one-factorizations of K_{10} all have different trains, although two ($F16$ and $F26$) have trains with the same indegree sequence. Dinitz (see [166]) tested 1,000,000 randomly generated one-factorizations of K_{12}, and found 905,461 different indegree sequences, and he obtained even better results for K_{14}. (It should also be realized that these samples may have included some isomorphic pairs.) No example has yet been found of non-isomorphic one-factorizations with isomorphic trains.

The tricolor vector does not behave as well as the train. Among the one-factorizations of K_{10} theer are 51 pairs, 8 triples and 2 sets of four pairwise non-isomorphic factorizations with the same vector ([77]). Interestingly, $F16$ and $F26$ have different vectors.

Exercises 11

11.1 Suppose $\mathcal{F}$ is a one-factorization of K_{2n}.

   (i) Show that if the automorphism group of $\mathcal{F}$ has order g, then there are $\frac{(2n)!}{g}$ one-factorizations isomorphic to $\mathcal{F}$.

   (ii) Find the automorphism groups of the six one-factorizations $\mathcal{F}_1$, $\mathcal{F}_2$, $\mathcal{F}_3$, $\mathcal{F}_4$, $\mathcal{F}_5$, $\mathcal{F}_6$ shown in Table 11.1; verify that their orders are 1344, 64, 16, 96, 24, and 42, respectively.

   (iii) Hence prove that there are 6240 one-factorizations of K_8.

11.2 Prove that the number of one-factorizations of K_{2n} is divisible by

$$\frac{(2n-3)!}{(n-2)!2^{n-2}}.$$

11.3 Prove that a one-factorization containing a d-division must have at least $2(d+1)$ vertices when d is odd, and at least $2(d+2)$ when d is even.

11.4 Prove Lemma 11.1.

11.5 Suppose a one-factorization $\mathcal{F}$ of K_{2n} has automorphism group of order 1. Prove that if $\mathcal{G}$ is the near-one-factorization of K_{2n-1} formed by deleting one vertex v from K_{2n} and from all members of $\mathcal{F}$, then $\mathcal{G}$ also has automorphism group of order 1. Is the converse true?

11.6 Verify that the one-factorization of K_{10} given at the end of the proof of Theorem 11.4 has no 3-division. What are its 2-divisions?

11.7 Prove that the staircase factorization of K_{2n} is isomorphic to the patterned factorization of K_{2n}.

11.8 Suppose $n = 2^{s-1}$. Prove that the geometric factorization $\mathcal{GG}_{2n}$ has train of length $2n - 1$.

11.9 A Latin square L of side $2n$, n odd, is constructed as follows. First, a symmetric idempotent Latin square of side n is constructed. Then in every case the entry i is replaced by the 2×2 array

$2i - 1$	$2i$
$2i$	$2i - 1$

Prove that the result is an intercalate square.

11.10 (*) Construct an intercalate square of order $2n$ for every even n except 2. [71]

11.11 Consider the lengths of trains of K_{2n}, when $2n > 6$. Prove or disprove the following:

   (i) there is no train of length 1;

   (ii) the only trains of length 2 are those of $\mathcal{G}K_{2n}$ where 3 and $2n - 1$ are coprime.

11.12 Prove that no one-factorization of K_{2n} has a train of length $2n - 2$. [164]

11.13 (?) Prove or disprove: if $n \geq 3$ then there is a one-factorization of K_{2n} with train of length t whenever $3 \leq t \leq 2n - 3$.

11.14 (?) Exhibit a pair of non-isomorphic one-factorizations with isomorphic trains, or prove that none can exist.

11.15 Prove that the non-isomorphic one-factorizations of K_8 have different tricolor numbers.

11.16 (*) Prove that the tricolor vector of $\mathcal{G}K_{2n}$ is given by:

   (i) $v_0 = 0$, $v_1 = \frac{2}{3}(2n - 1)(n - 1)(n - 3)$, $v_2 = (2n - 1)(n - 1)$, $v_i = 0$ for $i \geq 3$, if 3 does not divide $2n - 1$;

   (ii) $v_0 = 0$, $v_1 = \frac{2}{3}(2n - 1)(2n - 2)^2$, $v_2 = (2n - 1)(n - 2)$, $v_3 = 0$, $v_4 = \frac{1}{3}(2n - 1)$, $v_i = 0$ for $i \geq 5$, if 3 divides $2n - 1$. [77]

AUTOMORPHISMS AND ASYMPTOTIC NUMBERS OF ONE-FACTORIZATIONS

In this chapter we explore a few properties of automorphisms and automorphism groups of one-factorizations. We then proceed to prove that the number of isomorphism classes of one-factorizations of K_{2n} tends to infinity as n does. This result, which forms a nice extension of the result of Theorem 11.4, was proven independently in [11] and [107]; the proofs are quite similar, but [107] also includes another proof using quasigroups. Basically we shall follow [11].

First we consider the patterned one-factorization $\mathcal{P}_{2n}$; as usual we take the vertices to be $S = \{\infty\} \cup \mathbb{Z}_{2n-1}$, and write

$$P_i = \{(\infty, i)\} \cup \{(i - j, i + j) : 1 \leq j \leq n - 1\}.$$

So P_i consists of (∞, i) and all the edges (x, y) where $x + y = 2i$.

One obvious automorphism of $\mathcal{P}_{2n}$ is α_x ("add x"), defined by

$$\infty\alpha_x = \infty, \quad i\alpha_x = i + x,$$

where x is any member of $\mathbb{Z}_{2n-1}$. α_0 is the identity map i_S; when $x \neq 0$, α_x consists of a cycle of length $2n - 1$. It is interesting to observe that when $n > 3$ all automorphisms of $\mathcal{P}_{2n}$ must leave ∞ unchanged:

Theorem 12.1 (adapted from [78]) *Suppose $n \geq 4$. If π is an automorphism of the patterned factorizat ion $\mathcal{P}_{2n}$ on $S = \{\infty\} \cup \mathbb{Z}_{2n-1}$ then $\infty\pi = \infty$.*

Proof. Suppose $Aut(\mathcal{P}_{2n})$ contained a member π' such that $\infty\pi' \neq \infty$. Then $x\pi = \infty$ for some x, and $\pi = \alpha_x\pi'$ maps 0 to ∞. So it is sufficient to prove that no automorphism π exists for which $0\pi = \infty$.

Since $n \geq 4$, $\{1,2,3,4,5,6\}$ are distinct members of $\mathbb{Z}_{2n-1}$. If a, b, c and d are distinct members of $\mathbb{Z}_{2n-1}$ satisfying $a + b = c + d$, it follows that (a,b) and (c,d) are edges in the same factor P_i, where $2i = a + b$, and therefore $(a\pi, b\pi)$ and $(c\pi, d\pi)$ are in the same factor of $\mathcal{P}$, since π is an automorphism. Putting $d = 0$ we observe that this means

$$a + b = c \mapsto a\pi + b\pi = 2(c\pi)$$

whence

$$1\pi + 3\pi = 2(4\pi) \tag{12.1}$$
$$2\pi + 3\pi = 2(5\pi) \tag{12.2}$$
$$1\pi + 5\pi = 2(6\pi) \tag{12.3}$$
$$2\pi + 4\pi = 2(6\pi) \tag{12.4}$$

and $2(12.3) - 2(12.4) + (12.2) - (12.1)$ is

$$1\pi - 2\pi = 0$$

whence $1\pi = 2\pi$, which is impossible.   □

The condition that $n \geq 4$ is certainly necessary to the Theorem. When $n = 1$ or $n = 2$, any permutation is an automorphism of $\mathcal{P}_{2n}$. Determination of the automorphism group of $\mathcal{P}_6$ is left as an exercise, but we observe that $(\infty 0)(14)$ is one of the automorphisms, and it shifts ∞.

Corollary 12.1.1 *Suppose $n \geq 4$ and suppose π is an automorphism of $\mathcal{P}_{2n}$ which maps P_i to P_i and P_{i+1} to P_{i+1} for some i. Then π equals the identity map.*

Proof. Write the vertices of $\mathcal{P}_{2n}$ as $v_0, v_1, \cdots, v_{2n-1}$, where $v_0 = \infty$, and where P_i consists of the edges $(v_0, v_1), (v_2, v_3), \cdots, (v_{2n-2}, v_{2n-1})$, and P_{i+1} consists of $(v_1, v_2), (v_3, v_4), \cdots, (v_{2n-1}, v_0)$ — in particular $v_1 = i$ and $v_{2n-1} = i + 1$. (This is possible because $P_i \cup P_{i+1}$ is a Hamilton cycle — see Corollary 11.1.1.) Now it is an easy induction to show that $v_j \pi = v_j$ for every j: it is true for $j = 0$ by the Theorem; if it is proven for v_j and j is even, then $(v_j, v_{j+1}\pi)$ must be an edge in P_i, whence $v_{j+1}\pi = v_j$; and similarly for j odd.   □

We would say that the automorphism π of the Corollary "stabilizes P_i and P_{i+1}". As an extension of this idea, we shall say that an automorphism of a one-factorization $\mathcal{F}$ *stabilizes* $\mathcal{F}$ if it maps every factor in $\mathcal{F}$ to itself.

Corollary 12.1.2 *The only automorphism which stabilizes a patterned factorization other than* $\mathcal{P}_2$, $\mathcal{P}_4$ *or* $\mathcal{P}_6$ *is the identity.*

Proof. In the case of the patterned one-factorization, this follows from Corollary 12.1.1. For the patterned near-one-factorization, a stabilizing permutation must map the isolated vertex in each factor to itself. □

Lemma 12.2 *Suppose* $\mathcal{F}$ *is a one-factorization of* K_{2n} *with vertex-set* V, *and suppose* π *stabilizes* $\mathcal{F}$. *Then either* $\pi = i_V$ *or else* π *is the product of* n *2-cycles.*

Proof. Since edge (x, y) maps to $(x\pi, y\pi)$ under π, it follows that (x, y) and $(x\pi, y\pi)$ must lie in the same one-factor. In particular:

(i) if $x\pi = x$, then $y\pi = y$;

(ii) if $x\pi = y$, then $y\pi = x$.

Every possible edge occurs in K_{2n}, so (i) implies that if any vertex is fixed by π then π equals the identity. On the other hand, if $x\pi = y$ where $y \neq x$, then the disjoint cycle decomposition of π contains the 2-cycle (x, y). So any non-identity π satisfying the conditions is a product of n 2-cycles. □

Recall that Euler's totient function, denoted $\phi(n)$, is defined to equal the number of positive integers less than n which are relatively prime to n. (See, for example, [83].) It is well-known that if $p_1, p_2, \cdots, p_k$ are the distinct primes which divide n, then

$$\phi(n) = \frac{(p_1 - 1)(p_2 - 1) \cdots (p_k - 1)n}{p_1 p_2 \cdots p_k}.$$

Theorem 12.3 [11, 10] *If* $2n \geq 8$, *then* $Aut(\mathcal{P}_n)$ *has* $\frac{2n-1}{\phi(2n-1)}$ *elements.*

Proof. Suppose there are a_0 automorphisms of $\mathcal{P}_{2n}$ which take P_0 to itself. Then there will be a_0 automorphisms which carry P_0 to P_i: the permutations $\pi\alpha_i$, where π leaves P_0 invariant. So $Aut(\mathcal{P}_n)$ has $(2n - 1)a_0$ elements.

If x is any element of $\mathbb{Z}_{2n-1}$, define a permutation τ_x ("times x") by the laws $i\tau_x = ix$ when $i \in \mathbb{Z}_{2n-1}$, $\infty\tau_x = \tau_x$. Then τ_x is clearly an automorphism if and only if $\gcd(x, 2n - 1) = 1$. So we know $\phi(2n)$ such automorphisms. It is clear that they fix P_0 (more generally, $P_i\tau_x = P_{ix}$).

Now suppose π is any automorphism such that $P_0\pi = P_0$, and say $P_1\pi = P_i$. Since $P_0 \cup P_1$ is a Hamilton cycle, $P_0 \cup P_i$ must also be Hamilton. So, from Corollary 11.1.1, $\gcd(i, 2n-1) = 1$. Now $\pi(\tau_i)^{-1}$ is in $Aut(\mathcal{P}_{2n})$, and it maps P_0 to P_0 and P_1 to P_1. So, from Corollary 12.1.1, $\pi(\tau_i)^{-1}$ equals the identity, and $\pi = \tau_i$. Thus $a_0 = \phi(2n-1)$. So we have the result. $\square$

Corollary 12.3.1 *The automorphism group of the near-one-factorization derived from the patterned starter in $\mathbb{Z}_{2n-1}$ has $(2n-1)\phi(2n-1)$ elements.*

Proof. From Theorem 12.1, deletion of the symbol ∞ does not materially change the automorphisms. $\square$

An object whose only automorphism is the identity is called *rigid* or *asymmetric*. We know there is no rigid one-factorization on 2, 4, 6 or 8 points. However, we shall show that there is a rigid one-factorization of K_{2n} whenever $2n \geq 10$.

We shall use a special type of one-factorization which will also prove useful in later sections. Suppose X and Y are disjoint n-sets, where n is even, and $\mathcal{F}_X$ and $\mathcal{F}_Y$ are one-factorizations based on X and Y, respectively. Define $\mathcal{F}$ to be a one-factorization based on $X \cup Y$ in which $n-1$ of the factors are constructed by taking the union of a member of $\mathcal{F}_X$ with a member of $\mathcal{F}_Y$, and the other n factors form a one-factorization of the $K_{n,n}$ based on X and Y. Any such factorization is called a *twin factorization*.

When n is odd, a twin factorization is formed as follows. Take near-one-factorizations $\mathcal{F}_X$ and $\mathcal{F}_Y$ based on the n- sets X and Y and pair the factors in some way; n of the factors in $\mathcal{F}$ are formed by taking the union of the paired near-one-factors together with the edge joining the two isolated vertices. (We call this last-mentioned edge the twin-edge.) The n new edges formed in this way would together make up a one-factor of the $K_{n,n}$ based on X and Y. To complete $\mathcal{F}$, take factors which, together with the factor just mentioned, would form a one-factorization of the $K_{n,n}$.

In each case, the factors consisting entirely of edges joining X to Y will be called *cross* factors and the others *outside* factors. When n is odd every outside factor contains precisely one cross edge. $\mathcal{C}_{\mathcal{F}}$ and $\mathcal{O}_{\mathcal{F}}$ denote the sets of cross factors and outside factors in $\mathcal{F}$. If F is a factor in $\mathcal{O}_{\mathcal{F}}$ then F_X and F_Y denote the corresponding factors in $\mathcal{F}_X$ and $\mathcal{F}_Y$, respectively.

Lemma 12.4 [11] *Suppose $\mathcal{G}$ and $\mathcal{H}$ are twin factorizations of K_{2n}, where $2n \geq 10$, and suppose $\mathcal{G}_X$ arises from the patterned starter in a cyclic group. Then $\mathcal{O}_{\mathcal{G}}\pi = \mathcal{O}_{\mathcal{G}}$ and $\mathcal{C}_{\mathcal{G}}\pi = \mathcal{C}_{\mathcal{G}}$ for every π in $IG(\mathcal{G}, \mathcal{H})$.*

Proof. It will be convenient to write $n = 2r$ when n is even and $n = 2r + 1$ when n is odd. We assume that the twin factorizations $\mathcal{G}$ and $\mathcal{H}$ were defined using the same two n-sets X and Y of vertices. We assume the existence of a factor F in $\mathcal{O}_{\mathcal{G}}$ and a permutation π in $IG(\mathcal{G}, \mathcal{H})$ such that $F\pi$ is in $\mathcal{C}_{\mathcal{H}}$, and derive a contradiction. We write

$$
\begin{aligned}
T_X &= \{x : x \in X, x\pi \in Y\}, S_X = X\backslash T_X, \\
T_Y &= \{y : y \in Y, y\pi \in X\}, S_Y = Y\backslash T_Y.
\end{aligned}
$$

We say a factor E in $\mathcal{O}_{\mathcal{G}}$ is a *shifting factor* if $E\pi$ is in $\mathcal{C}_{\mathcal{G}}$, and a *still factor* if $E\pi$ is in $\mathcal{O}_{\mathcal{G}}$. An edge in (S_X, T_X) or (S_Y, T_Y) is a *shifting edge*.

Suppose first that n is even. There are r edges in F_X, and each edge contains one element of S_X and one element of T_X, so $|S_X| = |T_X| = r$. Similarly $|S_Y| = |T_Y| = r$. An edge belongs to a shifting factor if and only if it is a member of $(S_X, T_X) \cup (S_Y, T_Y)$; since there are exactly $2r^2$ such edges, there are exactly r shifting factors, and therefore $r - 1$ still factors, in $\mathcal{O}_{\mathcal{G}}$.

When n is odd there are, in F, r edges entirely in X, r entirely in Y, and one other, (x, y) say ($x \in X$, $y \in Y$). Let us assume x is in T_X. Then y is in T_Y and $|T_X| = |T_Y| = r + 1$, $|S_X| = |S_Y| = r$. (If x is in S_X, we can convert to an equivalent case with x in T_X by exchanging X and Y in the definition of $\mathcal{H}$.) The shifting factors contain r edges from each of (S_X, T_X) and (S_Y, T_Y). A still factor can contain at most one such edge. Say there are q shifting factors and therefore $(2r + 1) - q$ still factors; since F is a shifting factor, $0 \leq (2r + 1) - q \leq 2r$. Say s of the still factors contain shifting edges: $0 \leq s \leq (2r + 1) - q$. Counting shifting edges,

$$2qr + s = 2r(r + 1),$$

whence

$$q + \tfrac{s}{2r} = r + 1.$$

Given the bounds on s, either $q = r$ or $q = r + 1$. If $q = r$ we have $s = 2r$, which is too big; so $q = r + 1$ and $s = 0$. There are $r + 1$ shifting factors and r still factors, none of which contains a shifting edge. From this last fact we deduce that if r is even then every still factor contains $\tfrac{r}{2}$ edges from each of (S_X, S_X),

(T_X, T_X), (S_Y, S_Y) and (T_Y, T_Y), and the cross-edge joins T_X to T_Y; when r is odd, the cross-edge always joins S_X to S_Y.

Suppose G^1 and G^2 are two still factors in $\mathcal{O}_g$. If n is even then $G^1 \cup G^2$ contains no edge joining S_X to T_X, so G^1_X and G^2_X form a 2-division in $\mathcal{G}_X$. If n is odd, it is easiest to consider an extended factorization $\mathcal{E}_X$ based on $X \cup \{\infty\}$: define E^i to be formed from G^i_X by joining the isolated vertex in G^i to ∞. Then E^1 and E^2 again form a 2-division: if r is even then no edge joins S_X to $T_X \cup \{\infty\}$, and if r is odd then no edge joins $S_X \cup \{\infty\}$ to T_X.

Now $\mathcal{G}_X$ comes from the patterned starter. Without loss of generality label X as $\mathbb{Z}_{n-1} \cup \{\infty\}$ when n is even and as $\mathbb{Z}_n$ when n is odd; write G_i for the factor in $\mathcal{G}_X$ containing edge (∞, i) or isolated vertex i respectively, and in the latter case $E_i = G_i \cup \{(\infty, 0)\}$. From Corollary 11.1.1, both $G_i \cup G_{i+1}$ and $G_{i+1} \cup G_{i+2}$ form a Hamilton cycle in the case of n even and so do $E_i \cup E_{i+1}$ and $E_{i+1} \cup E_{i+2}$ when n is odd. So no more than one of $\{G_i, G_{i+1}, G_{i+2}\}$ can come from a still factor in $\mathcal{G}$, in either case. The subscripts can be taken modulo $n - 1$ or n, respectively. So $\mathcal{G}$ contains at most $\lfloor \frac{n-1}{3} \rfloor$ still factors, n even, and at most $\lfloor \frac{n}{3} \rfloor$ still factors, n odd. When n is even we have

$$r - 1 \leq \lfloor \tfrac{n-1}{3} \rfloor = \lfloor \tfrac{2r-1}{3} \rfloor,$$

whence $r \leq 2$ and $2n \leq 8$; when n is odd we have

$$r \leq \lfloor \tfrac{n}{3} \rfloor = \lfloor \tfrac{2r+1}{3} \rfloor$$

and again $r \leq 2$. So we have the result. $\square$

Lemma 12.5 *Suppose $\mathcal{G}$ and $\mathcal{H}$ are twin factorizations of K_{2n} on (X, Y), and π is a member of $IG(\mathcal{G}, \mathcal{H})$ which satisfies $\mathcal{O}_g \pi = \mathcal{O}_g$. Then either $x\pi \in X$ for all x in X or $x\pi \in Y$ for all x in X.*

Proof. This is easy when n is even, and the case $n = 1$ is trivial. So assume n to be odd, $n \geq 3$. If the Lemma is false, we can assume that X contains three elements, x, y and z, such that either $\{x\pi, y\pi\} \subseteq X$ and $z\pi \in Y$ or else $\{x\pi, y\pi\} \subseteq Y$ and $z\pi \in X$. In either case, both $(x\pi, z\pi)$ and $(y\pi, z\pi)$ are cross-edges; since (x, z) and (y, z) occur in members of $\mathcal{O}_g$, both $(x\pi, z\pi)$ and $(y\pi, z\pi)$ are cross-edges in members of $\mathcal{O}_\mathcal{H}$. But this means $z\pi$ occurs in two cross-edges in $\mathcal{O}_\mathcal{H}$, and by the construction of a twin factorization each vertex occurs in exactly one such edge. $\square$

We define a *special* twin factorization of K_{2n} to be a twin factorization on (X,Y) for which $\mathcal{F}_Y$ is rigid and $\mathcal{F}_X$ is the patterned factorization.

Lemma 12.6 *Suppose $\mathcal{G}$ and $\mathcal{H}$ are special twin factorizations on (X, Y), where $|X| = |Y| = n \geq 9$, and suppose π belongs to $IG(\mathcal{G}, \mathcal{H})$. Then the restrictions of π to X and Y map $\mathcal{G}_X$ to $\mathcal{H}_X$ and $\mathcal{G}_Y$ to $\mathcal{H}_Y$, respectively.*

Proof. From Lemma 12.4, $\mathcal{O}_\mathcal{G}\pi = \mathcal{O}_\mathcal{H}$, so from Lemma 12.5 either $X\pi = X$ or $X\pi = Y$. In the latter case $\mathcal{G}_X\pi = \mathcal{H}_Y$. But $\mathcal{H}_Y$ is rigid and $\mathcal{G}_X$ is not (by Theorem 12.3). So $X\pi = X$, whence $Y\pi = Y$ and we have the Lemma. $\square$

Corollary 12.6.1 *A special twin factorization is rigid.*

Proof. Take $\mathcal{G} = \mathcal{H}$ in the Lemma. Then $\mathcal{G}_Y = \mathcal{H}_Y$; and since this factorization is rigid, then only possible automorphism on $\mathcal{G}_Y$ is i_Y. So, given any automorphism π of $\mathcal{G}$ and any factor G in $\mathcal{O}_\mathcal{G}$, $G_Y\pi = G_Y$. So $G_X\pi = G_X$. That is, the restriction $\pi|X$ of π to X stabilizes $\mathcal{G}_X$. So, by Corollary 12.1.2, $\pi|X$ is the identity. So π equals the identity map. $\square$

Theorem 12.7 *There is a rigid one-factorization of K_{2n} whenever $2n \geq 10$.*

Proof. We observe that if $\mathcal{F}$ is a rigid one-factorization of K_{2n} then the result of deleting any vertex is a rigid near-one-factorization of K_{2n-1}. So the existence of a rigid one-factorization of $K_{2n}, 2n \geq 10$, not only implies that there is one for K_{4n} by Corollary 12.6.1 but also for K_{4n-2}. Therefore it is sufficient to exhibit rigid one-factorizations of K_{10}, K_{12}, K_{14} and K_{16}. This is done in Table 12.1 (the examples are from [11]). $\square$

Rigid factorizations are, in fact, quite common. For example, of the 396 one-factorizations of K_{10}, 298 are rigid [73]. However, as we stated above, there are no rigid one-factorizations when $2n < 10$ (see Exercise 11.1(ii)).

Theorem 12.8 *The number of isomorphism classes of rigid one-factorizations of K_{2n} goes to infinity with n.*

Proof. Let us write $R(2n)$ for the number of rigid one-factorizations of K_{2n}, and $R(2n-1)$ for the number of rigid near-one-factorizations of K_{2n-1}. Lemma 12.6 shows that if $\mathcal{G}$ and $\mathcal{H}$ are special twin factorizations which use non-isomorphic rigid one-factorizations (or near-one-factorizations) $\mathcal{G}_Y$ and $\mathcal{H}_Y$,

K_{10}: Use the factorization from the starter

$$\{\infty 012364857\}$$

but replace factors 0, 2 and 3 by

$\infty 0$	16	27	34	58
$\infty 2$	07	18	36	45
$\infty 3$	06	12	48	57

K_{12}: Use the patterned factorization,
but replace factors 0, 1 and 3 by

$\infty 0$	24	15	$3T$	67	89
$\infty 1$	02	38	49	56	$7T$
$\infty 3$	06	$1T$	29	47	58

K_{14}: Use the patterned factorization,
but replace factors 0, 1 and 3 by

$\infty 0$	15	24	$3T$	69	$7D$	$8E$
$\infty 1$	02	$3D$	$4E$	58	67	$9T$
$\infty 3$	06	$1D$	$2E$	49	$5T$	78

K_{16}: Use the factorization from the starter

$$\{\infty 0132E456A7T8D9B\}$$

but replace factors 0, 2 and 3 by

$\infty 0$	13	$2E$	46	$5B$	78	$9D$	TA
$\infty 2$	$0E$	19	35	$4A$	67	$8D$	TB
$\infty 3$	08	$1E$	$2D$	45	$6A$	$7T$	$9B$

Table 12.1 Rigid one-factorizations. For convenience, $T\ E\ D\ A\ B$ stand for
10 11 12 13 14.

then $\mathcal{G}$ and $\mathcal{H}$ are not isomorphic. So, if we prove that we can produce c_n non-isomorphic twin factorizations from the same $\mathcal{G}_Y$, we will have $c_n R(n)$ non-isomorphic special twin factorizations, and

$$R(2n) \geq c_n R(n).$$

Let $\mathcal{G}_Y$ be any rigid one-factorization of K_{2n}, and $\mathcal{G}_X$ be a copy of $\mathcal{P}_{2r}$. Then there are $(2r-1)!$ different ways of putting the factors of $\mathcal{G}_X$ together with those of $\mathcal{G}_Y$ to form outside factors. Putting each of these with some fixed one-factorization of $K_{2r,2r}$ gives $(2r-1)!$ different special twin factorizations.

Now suppose two of those factorizations, $\mathcal{G}$ and $\mathcal{H}$ say, were isomorphic, and suppose π is an isomorphism such that $\mathcal{G}\pi = \mathcal{H}$. From Lemma 12.6, $\mathcal{G}_X\pi = \mathcal{H}_X$, and since $\mathcal{G}_X$ and $\mathcal{H}_X$ are just copies of $\mathcal{P}_{2r}$ (with the factors ordered differently), π is an automorphism of $\mathcal{P}_{2r}$. So if the $(2r-1)!$ factorizations are partitioned into isomorphism classes, there are at most $|Aut(\mathcal{P}_{2r})| = \frac{2r-1}{\phi(2r-1)}$ members in each class. Therefore

$$R(4r) \geq \frac{(2r-2)!}{\phi(2r-1)}R(2r)$$
$$\geq (2r-3)!R(2r). \tag{12.5}$$

(Of course the number is considerably bigger — different sets of cross factors give non-isomorphic factorizations.)

In the same way, when $n = 2r+1$, we obtain

$$R(4r+2) \geq \frac{(2r)!}{\phi(2r)}R(2r+1)$$
$$\geq (2r-1)!R(2r+1). \tag{12.6}$$

Finally, we need to derive bounds on $R(4r-1)$ and $R(4r+1)$, in order to use these inequalities recursively. But suppose we write $\mathcal{F}_x$ for the near-one-factorization derived from $\mathcal{F}$ by deleting the symbol x. If $\mathcal{F}$ and $\mathcal{G}$ are not isomorphic then clearly $\mathcal{F}_x$ and $\mathcal{G}_y$ are not isomorphic (if $\mathcal{F}_x\pi = \mathcal{G}_y$, define $x\pi = y$ and π becomes an isomorphism from $\mathcal{F}$ to $\mathcal{G}$), while if $\mathcal{F}$ is rigid then $\mathcal{F}_x$ and $\mathcal{F}_y$ are non-isomorphic (if $\mathcal{F}_x\pi = \mathcal{F}_y$ then one defines $x\pi = y$ to define a non-trivial automorphism of $\mathcal{F}$.) Of course, $\mathcal{F}_x$ is rigid if $\mathcal{F}$ is. So

$$R(2n-1) \geq 2nR(2n). \tag{12.7}$$

This together with (12.5) and (12.6) proves that $R(2n)$ (and $R(2n-1)$) goes to infinity with n. $\qquad\square$

Corollary 12.8.1 *The number of isomorphism classes of one-factorizations of K_{2n} goes to infinity with n.* $\qquad\square$

Exercises 12

12.1 Find the number of automorphisms of $\mathcal{P}_6$.

12.2 Prove that Corollary 12.1.1 does not hold of the condition "$n \geq 4$" is relaxed.

12.3 Suppose $\mathcal{G}$ and $\mathcal{H}$ are twin factorizations of K_{2n} and π is an isomorphism from $\mathcal{G}$ to $\mathcal{H}$. Prove that if $2n \equiv 2$ or $4 \pmod 8$ then $\mathcal{O}_{\mathcal{G}}\pi = \mathcal{O}_{\mathcal{H}}$ and $\mathcal{C}_{\mathcal{G}}\pi = \mathcal{C}_{\mathcal{H}}$.

12.4 Prove Lemma 12.5 in the case when n is even.

12.5 Check that the factorizations in Table 12.1 are rigid.

12.6 By considering cross factors, refine equation (12.5) to

$$R(4r) \geq \frac{(2r-1)!(2r-2)!}{\phi(2r-1)} R(2r)$$

and deduce a similar result for $R(4r+2)$.

12.7 Prove that the one-factorization of K_{10} given at the end of Chapter 11 has automorphism group of size 2, and that the non-trivial automorphism is $(07)(18)(26)(35)(49)$. Hence prove that the near-one-factorizations derived by deleting one vertex from this factorization are all rigid. Deduce that

$$R(9) > 10R(10)$$

(in other words, the inequality (12.7) is not best-possible).

SYSTEMS OF DISTINCT
REPRESENTATIVES

We digress from the subject of one-factorizations to present three theorems of combinatorial set theory. The reader may wish to skip the proofs, but the results will be used in later chapters.

By a *set system S* we mean a list $(S_1, S_2, \cdots, S_r)$ of sets. A *system of distinct representatives* or SDR for S is a way of selecting a member x_i from each S_i in such a way that $x_1, x_2, \cdots, x_r$ are all different. For example, consider the sets

$$123, \quad 124, \quad 134, \quad 235, \quad 246, \quad 1256.$$

One system of distinct representatives for them is

$$1, \ 2, \ 3, \ 5, \ 4, \ 6.$$

There are several others. On the other hand, the sets

$$124, \quad 124, \quad 134, \quad 23, \quad 24, \quad 1256$$

have no system of distinct representatives.

Theorem 13.1 [81] *Suppose* $S = \{S_1, S_2, \cdots, S_r\}$ *is a system of sets, and any union of k members of S contains at least k elements, for every k, $1 \leq k \leq r$. Then S has a system of distinct representatives.*

Proof. We proceed by induction on the number of sets. If S has one member, the result is obviously true. Assume the Theorem to be true for $r = 1, 2, \cdots, n-1$. Suppose that S has n members, $S_1, S_2, \cdots, S_n$, and that any union of k of the sets has size at least k, for $1 \le k \le n$. By induction, any collection of k of the sets has a system of distinct representatives, for $k < n$. We distinguish two cases.

(i) Suppose that no union of k of the S_i contains fewer than $k+1$ elements, for any $k < n$. Select any element x_1 of S_1, and write $T_i = S_i \backslash \{x_i\}$, for $1 \le i \le n$. Then the union of any k of the T_i has at least k elements, for $k = 1, 2, \cdots, n-1$. So, by the induction hypotheses, there exists a system of distinct representatives for $\{T_2, T_3, \cdots, T_n\}$, say $(x_2, x_3, \cdots, x_n)$. Then $(x_1, x_2, \cdots, x_n)$ is a system of distinct representatives for S.

(ii) Suppose some union of k sets has size k: without loss of generality, suppose

$$|S_1 \cup S_2 \cup \cdots \cup S_k| = k.$$

For each i greater than k, write T_i to mean S_i with all elements of $S_1 \cup S_2 \cup \cdots \cup S_k$ deleted. It is easy to see that $T_{k+1}, T_{k+2}, \cdots, T_n$ satisfy the conditions of the Theorem, because if some h of them had fewer than h elements in their union, say $|T_{k+i_1} \cup T_{k+i_2} \cup \cdots \cup T_{k+i_h}| < h$, then

$$|S_1 \cup S_2 \cup \cdots \cup S_k \cup S_{k+i_1} \cup \cdots \cup S_{k+i_h}| < h + k,$$

contradicting the induction hypothesis. So we can find a system of distinct representatives $(x_{k+1}, x_{k+2}, \cdots, x_n)$ for $T_{k+1}, T_{k+2}, \cdots, T_n$, and they will also be a system of distinct representatives for $S_{k+1}, S_{k+2}, \cdots, S_n$. We can also find a system of distinct representatives for $S_1, S_2, \cdots, S_k$. It is clear that these two systems are disjoint, and the union is a system of distinct representatives for S. $\qquad\qquad\square$

Theorem 13.2 [112] *Suppose S satisfies the conditions of Theorem 13.1, and further suppose the set M is such that $|M \cap S_i| \le t$ for every i, and every element of M belongs to at least t members of S. Then S has a system of distinct representatives which contains every member of M.*

The elements of M are called *marginal elements* for the set system S.

Proof. Among systems of distinct representatives for S, let X be one which contains the maximum number of elements of M. We wish to show that X contains all of M. We assume to the contrary that X does not contain all elements of M, and derive a contradiction. Write x_i for the representative of S_i in X.

We define a *chain* $(S_{j_1}, S_{j_2}, \cdots, S_{j_a})$ to be an ordered collection of sets in S such that x_{j_i} belongs to $S_{j_{i+1}}$ when $1 \leq i < a$. We would call the above example a chain *from* S_{j_1} *to* S_{j_a}, and say S_{j_a} *follows* S_{j_1}. If T follows S and U follows T, then clearly U follows S. Notice that each set in S is a (one-element) chain, so each set follows itself.

Suppose d is a marginal element which does not belong to X. We prove that if d is in S_j then there is a chain from S_j to S_k, for some k such that x_k is not in M. For, suppose not. Let $S_{i_1}, S_{i_2}, \cdots, s_{i_a}$ be the sets in S which follows S_j. Then $x_{i_1}, x_{i_2}, \cdots, x_{i_a}$ all belong to M, and each occurs at least t times among $\{S_{i_1}, S_{i_2}, \cdots, S_{i_a}\}$, because each set in S containing x_{i_b} follows S_{i_b}, so it follows S_j. Therefore $\{S_{i_1}, S_{i_2}, \cdots, S_{i_a}\}$ contain at least $at + 1$ copies of marginal elements (each of the a representatives of the sets occurs t times, and d, which is not a representative, occurs at least once, in S_j). So one of the a sets contains more than t marginal elements, a contradiction. So one of the S_{i_k} must have a non-marginal representative.

Now let d be a marginal element which is not in X, S_j a set containing d, and $(S_j = S_{i_1}, S_{i_2}, \cdots, S_{i_a})$ the shortest chain such that X_{i_a} is not in M. Then $x_j, x_{i_2}, \cdots, x_{i_{a-1}}$ are in M. We construct a new system of distinct representatives $Y = \{y_i\}$, where y_i represents S_i, as follows: $y_j = d$, $y_{i_2} = x_j$, $y_{i_3} = x_{i_2}, \cdots, y_{i_a} = x_{i_{a-1}}$, and $y_i = x_i$ otherwise. Then Y contains more elements of M than X does -- all of $M \cap X$, together with d. This contradicts the maximality of X. So the selection of d must have been impossible, and $M \subseteq X$, as required. $\qquad\square$

Theorem 13.3 (Bondy, see [130]) *Suppose X and Y are sets of sizes m and n respectively, where $m \leq n$. Let $\{Y_1, Y_2, \cdots, Y_n\}$ be any collection of m-subsets of Y such that every member y of Y is contained in exactly m of the Y_i. Then the $K_{m,n}$ based on vertex-sets X and Y has a factorization $\{M_1, M_2, \cdots, M_n\}$ where M_i is a perfect matching between X and Y_i.*

Proof. Let us call a system of sets m-*uniform* if each set has m elements and each element of the union of the sets occurs in exactly m sets. Let T be a set of k members of a uniform set the union of whose members is $U = \{u_1, u_2, \cdots, u_s\}$, and say u_i belongs to r_i members of T. Then $r_i \leq m$ for each i, so $\Sigma r_i \leq ms$. On the other hand, Σr_i counts all elements of each element of T, so $\Sigma r_i = mk$. So $k \leq s$, and U contains at least k elements. So from Theorem 13.1, any m-uniform system has a system of distinct representatives.

Now $\{Y_1, Y_2, \cdots, Y_n\}$ is m-uniform. Let $S_1 = (y_{11}, y_{12}, \cdots, y_{1n})$ be a system of distinct representatives, where $y_{1,j} \in Y_j$. Then $\{y_{i1}, y_{i2}, \cdots, y_{in}\}$ is just Y in some order. So if $Y_{1j} = Y_j \backslash \{y_{ij}\}$, then the system $\{Y_{i1}, Y_{i2}, \cdots, Y_{in}\}$ is $(m-1)$-uniform, and it also contains a system of distinct representatives, say $S_2 = (y_{21}, y_{22}, \cdots, y_{2n})$. In this way we may select m systems of distinct representatives, $S_1, S_2, \cdots, S_m$, where $S_i = (\cdots, y_{ij}, \cdots)$. For each j, $Y_j = \{y_{1j}, y_{2j}, \cdots, y_{mj}\}$ in some order. Because members of each system of distinct representatives are eliminated from the set which they represented before the next system of distinct representatives is constructed, $Y_{ij} \neq Y_{hj}$ unless $i = h$.

We now define M_j to be the set of edges

$$M_j = \{x_i j_{ij} : 1 \leq i \leq m\}.$$

M_J will be a matching between X and Y_j, and it is clear that the different M_j are edge-disjoint. So $M_1, M_2, \cdots, M_n$ constitute the required factorization. $\square$

Exercises 13

13.1 Suppose the set system $\mathcal{S}$, which contains r sets, has a system of distinct representatives, and each set in $\mathcal{S}$ has at least t elements. Prove that if $t \leq r$ then $\mathcal{S}$ has at least $t!$ different systems of distinct representatives.

13.2 A set of tasks is to be performed by a set of workers. It is required to assign each task to a different worker; no worker can be assigned to a task for which she is not qualified. Prove that a necessary and sufficient condition that all tasks can be assigned is that for any set of r tasks, there is a set of at least r workers such that each is capable of performing at least one of the tasks.

13.3 M is a square matrix of zeros and ones with every row and column sum equal to r. Prove that there exist permutation matrices $P_1, P_2, \cdots, P_r$ such that

$$M = P_1 + P_2 + \cdots + P_r.$$

SUBFACTORIZATIONS AND ASYMPTOTIC NUMBERS OF ONE-FACTORIZATIONS

Consider a graph G with a one-factorization $\mathcal{F} = \{F_1, F_2, \cdots\}$. Suppose H is a subgraph of G. Normally $H \cap F_i$ will consist of some edges and some vertices. It may be, however, that for every i, $H \cap F_i$ consists either only of edges or only of vertices; in other words each $H \cap F_i$ either is a one-factor of H or has edge-set $\emptyset$. If this occurs, order the F_i so that $H \cap F_1, H \cap F_2, \cdots, H \cap F_t$ are one-factors in H, while $H \cap F_j$ has no edges for $j > t$. Then the first t factors $H \cap F_j$ form a one-factorization of H, and this is called a *subfactorization* of $\mathcal{F}$ (or a *subfactorization of H in G*).

We are basically interested in the case where both H and G are complete. Suppose $\mathcal{F}$ is a one-factorization of K_{2n} which contains a one-factorization $\mathcal{E}$ of K_{2s} as a subfactorization. Without loss of generality, assume the K_{2s} and K_{2n} to be based on $\{0, 1, \cdots, 2s - 1\}$ and $\{0, 1, \cdots, 2n - 1\}$, and assume F_i contains E_i for $1 \le i \le 2s - 1$. If $k \ge 2s$, then each of $0, 1, \cdots, 2s - 1$ must appear in a different edge of F_k, so F_k contains at least $2s$ edges. Therefore $2n \ge 4s$.

To prove the converse we follow the method used by Cruse [43] to obtain a slightly more general result. Suppose $P = (p_{ij})$ is an $r \times r$ array with entries from $\{1, 2, \cdots, v\}$. Write $N_P(i)$ for the number of times symbol i appears in the array P. Then P is called a *proper (v, r)-array* if

(i) no entry is repeated in any row or column of P;

(ii) $p_{ij} = p_{ji}$ for $1 \le i, j \le r$;

(iii) $N_P(i) \ge 2r - v$ for $1 \le i \le v$;

(iv) $N_P(i) \equiv v \pmod 2$ for at least r of the values of i, $1 \le i \le v$.

In particular, a proper (v, v)-array is clearly a symmetric Latin square of side v.

Lemma 14.1 [43] *Suppose P is a proper (v, r)-array, where $r < v$. Then there exists a proper $(v, r+1)$-array Q such that $p_{ij} = q_{ij}$ when $1 \le i, j \le r$.*

Proof. Suppose P is a proper (v, r)-array. We define several subsets of $\{1, 2, \cdots, v\}$ associated with P, as follows:

$$
\begin{aligned}
S_j &= \{i : i \text{ does not appear in row } j\}; \\
S_0 &= \{i : N_P(i) \not\equiv v(\text{mod } 2)\}; \\
D &= \{i : N(i) = 2r - v; \}; \\
E &= \{i : N(i) = 2r - v + 1\}; \\
F &= \{i : N(i) > 2r - v + 1\}; \\
M &= D \cup E.
\end{aligned}
$$

We denote $|S_0|$, $|D|$, $|E|$, $|F|$ and $|M|$ by s, d, e, f and m respectively. Then $m = d + e$ and $v = d + e + f = m + f$. Clearly $|S_j| = v - r$ for every r. Moreover property (iv) tells that $s \le v - r$; since $E \subseteq S_0$, we have $e \le v - r$.

Each member of D occurs $2r - v$ times in R, each member of E occurs $2r - v + 1$ times and each member of F at most r times, so R has at most

$$
\begin{aligned}
& (2r - n)d + (2r - n + 1)e + rf \\
= \ & (2r - n)m + e + r(n - m) \\
\le \ & (r - n)m + n - r + rn
\end{aligned}
$$

entries. But R has r^2 entries, so

$$
(r - n)m + (n - r) + rn \ge r^2;
$$

simplifying,

$$
(n - r)m \le (n - r)(r + 1)
$$

so $m \le r + 1$. Moreover, if $s = 0$, whence $e = 0$, we obtain

$$
(r - n)m + rn \ge r^2,
$$

and $m \le r$ in that case.

We wish to apply Theorem 13.2 to the system of sets $\mathcal{S} = \{S_0, S_1, \cdots, S_r\}$ and marginal set M. We first observe that none of $S_0, S_1, \cdots, S_r$ contains more than $v - r$ entries, so certainly none contains more than $v - r$ members of M. Moreover, if $N(i) = 2r - v$, then i belongs to $r - (2r - v) = v - r$ of the sets $S_1, S_2, \cdots, S_r$, while if $N(i) = 2r - v + 1$ then i belongs to $v - r - 1$ of those sets and also to S_0; in either case, i belongs to at least $v - r$ members of S.

Next we prove that, unless S_0 is empty, any union of k members of S contains at least k elements. For suppose such a union contained at most $k - 1$ different elements. Since no member of $\{1, 2, \cdots, v\}$ can belong to more than $v - r$ sets, by (iii), we find the sum of the numbers of elements in the k sets is at most $(k - 1)(v - r)$. But S_0 has at least one element, and the other $k - 1$ sets in the union will have at least $v - r$ elements, giving a sum of at least $(k-1)(v-r)+1$. In a similar way we see that even if S_0 is empty, any union of k of $S_1, S_2, \cdots, S_r$ contains at least k elements.

Combining these results, we see that if S_0 is non-empty we can find a system $(s_0, s_1, \cdots, s_r)$ of distinct representatives for $S_0, S_1, \cdots, S_r$, and if S_0 is empty we can find a system of distinct representatives $s_1, s_2, \cdots, s_r$ for $S_1, S_2, \cdots, S_r$. In each case s_i is the representative of S_i, and every member of M occurs among the representatives.

We form an array Q as follows. If i and j are less than $r + 1$, $q_{ij} = p_{ij}$; $q_{j,r+1} = q_{r+1,j} = s_j$. We take $q_{r+1,r+1} = s_0$ if S_0 was not empty; if S_0 was empty, any element not already in the system of distinct representatives may be used. Properties (i) and (ii) obviously hold. If $N_P(i) = 2r - v$ then i belongs to M but not to S_0, so $i = s_j$ for some $j > 0$, and $N_Q(i) = 2r - v + 2$, while if $N_P(i) = 2r - v + 1$ then either $i = s_0$ and $N_Q(i) = 2r - v + 2$ or else $N_Q(i) = 2r - v + 3$, so $N_Q(i) \geq 2(r+1) - v$ for every i. And $N_Q(i) \equiv v \pmod 2$ for at least $r + 1$ of the values i: if S_0 is not empty then $N_Q(i) \equiv v \pmod 2$ whenever $N_P(i) \equiv v \pmod 2$, and also when $i = s_0$; while if S_0 is empty, $N_Q(i) \equiv v \pmod 2$ for all values except $i = q_{r+1,r+1}$, which is $v - 1$ values. This shows that Q is a proper $(v, r+1)$-array except possibly in the case where $v = r + 1$; but in that case Q is a symmetric Latin square of side n and every entry appears n times. □

By the obvious induction we can prove:

Corollary 14.1.1 *A proper (v, r)-array can be embedded in a symmetric Latin square of side v.* □

Now suppose r and v are odd positive integers, say $r = 2s - 1$ and $v = 2n - 1$. Say $\mathcal{E}$ is a one-factorization of K_{2s}, and P is the symmetric Latin square of side r which is equivalent to $\mathcal{E}$. Then P can be embedded in a symmetric Latin square Q of side v. Reorder the entries of Q so as to achieve an idempotent square (only the elements greater than r need to be changed). If $\mathcal{F}$ is the one-factorization equivalent to this Latin square, then $\mathcal{E}$ is a subfactorization of $\mathcal{F}$. We have proven:

Theorem 14.2 *There exists a subfactorization of K_{2s} in K_{2n} if and only if* $2n \geq 4s$. □

In order to construct subfactorizations we use two related ideas from design theory: transversal designs and frames.

A *transversal design* $TD(k, n)$ consists of a finite set of kn elements called *points* which are partitioned into k subsets called *groups* of size n, together with a collection of k-subsets of the points called *blocks*, with the property that two points belong either to precisely one group or to precisely one block, but not both. Clearly there will be n^2 blocks. It is easy to prove that a $TD(k, n)$ exists if and only if there is a set of $k - 2$ pairwise orthogonal Latin squares of size n.

Frames were defined as two-dimensional objects for use in the construction of two-dimensional arrays such as Room squares. We shall adapt the definition to the simpler case of one-factorizations. Suppose T is the set of integers $\{1, 2, \cdots, t\}$, and $\{S_1, S_2, \cdots, S_k\}$ is some partition of T, where S_j has s_j elements. We define an $(S_1, S_2, \cdots, S_k)$-*semiframe* to be a set of partial one-factors $F_1, F_2, \cdots, F_t$ of the complete graph K_t based on T. If i belongs to S_j, then F_i contains every element of T except the elements of S_j, once each. The totality of all pairs includes every pair $\{x, y\}$ in which x and y belong to different sets S, precisely once. The multiset $\{s_1, s_2, \cdots, s_k\}$ is called the *type* of the semiframe; we write "type $1^\alpha 3^\beta \cdots$" to mean there are α sets of size 1, β of size 3, $\cdots$ in the partition. The sets S_j are called *holes*, and the set of factors $\{F_i : i \in S_j\}$ is also called a *hole of size s_j*.

To define a frame, one asks that the pairs in a semiframe be distributed in a two-dimensional array where the rows correspond to factors and the columns satisfy restrictions similar to the rows. (This is why the word "hole" is used: it corresponds to an empty subarray of size $s_j \times s_j$.) Frames are more restrictive than semiframes. In Table 14.1 we give several examples of semiframes; of these, only type $1^8 3^1$ can be realized as a frame. For further material on frames, see for example [142], [143] and [54].

type $1^4 3^1$

123		45	67
		46	57
		47	56
4	15	26	37
5	16	27	34
6	17	24	35
7	14	25	36

type 1^5

1	23	45
2	14	35
3	15	24
4	13	25
5	12	34

type 2^4

12	36	47	58
	38	45	67
34	16	28	57
	17	25	68
56	18	24	37
	13	27	48
78	14	26	35
	15	23	46

type $1^6 3^1$

123		47	59	68
		48	57	69
		49	58	67
4	17	29	38	56
5	18	27	39	46
6	19	28	37	45
7	14	26	35	89
8	15	24	36	79
9	16	25	34	78

type $1^8 3^1$

123		48	59	6T	7E
		49	58	6E	7T
		4T	5E	68	79
4	15	26	3T	78	9E
5	14	27	3E	69	8T
6	17	24	35	89	TE
7	16	25	34	9T	8E
8	19	2T	36	4E	57
9	18	2E	37	46	5T
T	1E	28	39	45	67
E	1T	29	38	47	56

Table 14.1 Some semiframes.

We shall call a semiframe *odd* if t is odd (and consequently all the s_j are odd); otherwise it is *even*. We shall use odd semiframes. Any odd semiframe of side t can be converted into a one-factorization of K_{t+1} by replacing the hole corresponding to S_j by a one-factorization of the K_{s_j+1} based on $\{\infty\} \cup S_j$, so odd semiframes correspond to one-factorizations with certain collections of subfactorizations. (Again the situation is less straightforward in the case of frames, because of the nonexistence of Room squares of sides 3 and 5.) In particular a semiframe of type 1^k is just a near-one-factorization of K_k, and exists for all odd k; while Theorem 14.2 guarantees semiframes of type $1^{i+j}i^1$ for all odd i and j.

Several constructions for semiframes are available. We give one as an example. It is a simplification of a frame construction in [54].

Lemma 14.3 *Suppose k, n and r are odd positive integers, and $0 \leq t \leq n$. If there exist a $TD(k,n)$ and a semiframe of type $1^{k-1}r^1$, then there is a semiframe of type $n^{k-1}(n + (r-1)t)^1$.*

Proof. Suppose the $TD(k,n)$ has groups $G_1, G_2, \cdots, G_k$. Select a t-subset W of G_k. If x is in W, write $S(x)$ to denote an r-set $\{x_1, x_2, \cdots, x_r\}$ of new points; for other points x it will sometimes be convenient to let $S(x)$ denote x. Write $G_k^* = \bigcup_{x \in G_k} S(x)$. For each block $A = \{x^1, x^2, \cdots, x^k\}$ of the $TD(k,n)$, where x^i lies in G_i, select any semiframe $\mathcal{F}_A$ whose holes are the sets $\{x^i\}, 1 \leq i < k$, and $S(x^k)$. $\mathcal{F}_A$ will be of type 1^k (if x is not a member of W) or $1^{k-1}r^1$, both of which exist. Now define a semiframe $\mathcal{F}$ as follows: if $S(x)$ is a singleton then a factor F_x is constructed by taking the union of the factors with hole $\{x\}$ from all $\mathcal{F}_A$ where A contains x; if $S(x)$ is an r-set then for each block A containing x, the r factors corresponding to hole $S(x)$ in $\mathcal{F}_A$ are ordered arbitrarily as $F_{A,1}, F_{A,2}, \cdots, F_{A,r}$ say and r new factors $F_{x,1}, F_{x,2}, \cdots, F_{x,r}$ are formed by defining $F_{x,i}$ to be the union of all the $F_{A,i}$. $\mathcal{F}$ consists of all the F_x and $F_{x,i}$. By the definition of a transversal design, the pair $\{x, y\}$ occurs together in one block if x and y are in different groups, and none otherwise, so the edges in the one-factors are the appropriate ones. It is clear that all the factors corresponding to a hole in G_i contain all the points outside G_i once, and none of the points of G_i, for $1 \leq i \leq k - 1$, and similarly for G_k^*. So $\mathcal{F}$ is a semiframe with holes $G_1, G_2, \cdots, G_{k-1}, G_k^*$. It clearly is of the required type. $\square$

As an application of this result, we prove (again!) that the number of one-factorizations goes to infinity. We can restrict ourselves to the case of the Lemma when $r = 3$ and $k = 5$. The building-blocks are a $TD(5, n)$, which exists for all $n > 10$ (see Theorem 4.4), and semiframes of types 1^5 and $1^4 3^1$. There are six different (not necessarily nonisomorphic) semiframes of type 1^5, since they are equivalent to one-factorizations of K_6. There are at least 420 different semiframes of type $1^4 3^1$: in the example in Table 14.1, the 3-set corresponding to the hole could be chosen in 35 different ways; and, once it is chosen, one could complete the semiframe in twelve ways.

Lemma 14.4 *When $n > 10$ and $0 \leq t \leq n$ there are at least*

$$6^{(n-t)n} \cdot 420^n \cdot 2^5$$

different one-factorizations of $K_{5n+2t+1}$ on a given symbol set.

Proof. Consider the construction used in Lemma 14.3. There were $(n - t)n$ blocks which contained no member of W; in each of those the semiframe required is of type 1^5, which can be chosen in six ways. The remaining n blocks receive a type $1^4 3^1$. So there are $6^{(n-t)n} \cdot 420^n$ semiframes of type $n^4(n + 2t)^1$.

Each semiframe can be built into a one-factorization by inserting four one-factorizations of K_{n+1} and one one-factorization of K_{n+2t+1}. There are two non-isomorphic one-factorizations of K_{n+1}, by Theorem 11.4, and two of K_{n+2t+1}. So we have 2^5 ways of forming a one-factorization of $K_{5n+2t+1}$. □

Theorem 14.5 *There are at least*

$$\frac{6^{(n-t)n} \cdot 420n \cdot 2^5}{(5n + 2t + 1)!}$$

non-isomorphic one-factorizations of $K_{5n+2t+1}$, when $n > 10$, n is odd and $0 \leq t \leq n$.

Proof. If two one-factorizations are isomorphic then one can be obtained from the other by a permutation of the symbols. So there can be at most $(5n+2t+1)!$ one-factorizations of $K_{5n+2t+1}$ isomorphic to a given one. Dividing this number into the lower bound of Lemma 14.4 gives the stated bound on the number of isomorphism classes. □

Corollary 14.5.1 *The number of one-factorizations of a complete graph goes to infinity with the order of the graph.*

Proof. The above bound can be simplified by replacing t by $n-1$ in the denominator and by 0 in the numerator. Then, when $5n < v < 7n$, K_v has more than

$$6^{n^2} \cdot 2^5 \cdot (7n)!^{-1}$$

one-factorizations. One version of Stirling's formula says $s! = (2\pi s)^{\frac{1}{2}}(\frac{s}{e})^s e^\alpha$ for some α lying between $(12s+1)^{-1}$ and $(12s)^{-1}$. Putting $s = 7n$, the bound is at least

$$6^{n^2} \cdot 2^5 \cdot (14\pi n)^{-\frac{1}{2}} \cdot (7n)^{-7n} \cdot e^{7n} \cdot e^{-\alpha}$$

which is a constant multiple of

$$6^{n^2} \cdot (7n)^{-7n-\frac{1}{2}} \cdot e^{-\alpha}$$

Since $e^{-\alpha} > 1$ and $(7n)^{-7n-\frac{1}{2}} > n^{-8n}$ for all large n, we have a lower bound of

$$(6^n e^7 n^{-1})^n,$$

and $6^n n^{-1}$ approaches infinity with n. $\qquad\square$

The above proof is a simplification of the corresponding proof for Room squares in [54]. Another proof, also proceeding by dividing a lower bound on the number of different one-factorizations by the maximum size of an isomorphism class, appears in [36].

Exercises 14

14.1 In the notation of Lemma 14.1, prove that if $r = v - 1$ then S_0 cannot be empty.

14.2 Prove that a $TD(k, n)$ exists if and only if there exist $k - 2$ pairwise orthogonal Latin squares of order n.

14.3 Construct semiframes of types 2^3 and $1^3 3^2$.

14.4 Verify that there are at least 420 semiframes of type $1^4 3^1$.

14.5 How large must n be in order that

$$(7n)^{-7n-\frac{1}{2}} > n^{-8n}?$$

15

CYCLIC ONE-FACTORIZATIONS

We say a one-factorization $\mathcal{F}$ of K_{2n} is *cyclic* if there exists an automorphism α of $\mathcal{F}$ which acts on the vertices of K_{2n} as a cycle of length $2n$. Cyclic one-factorizations were introduced in [85].

In many structural problems of graph theory or combinatorial design theory, it is fruitful to consider cyclic automorphisms. In a one-factorization of K_{2n}, any automorphism acts naturally on two sets, the $2n$ vertices and the $2n - 1$ edges. For this reason more attention has been paid to cyclic automorphisms of order $2n - 1$. (One can have a cycle of order $2n - 1$ which acts on the $2n$ vertices by fixing one of them.) However, cyclic one-factorizations have some interesting properties.

Suppose $\mathcal{F}$ is a cyclic one-factorization of K_{2n}. It will be convenient to write the vertices of K_{2n} as the integers modulo $2n$, and to define $|x|$ by

$$|x| = \begin{cases} x & \text{if } 0 \leq x < n, \\ 2n - x & \text{otherwise.} \end{cases}$$

Since the modulus $2n$ is even, parity is preserved under addition mod $2n$, and we can speak of a vertex being odd or even. The edge xy is odd or even according as $|x - y|$ is. Without loss of generality, we may assume α to be the map

$$\alpha(x) = x + 1.$$

Let us consider the action of a α on the set of edges of K_{2n}. Two edges ab and cd will lie in the same orbit under α if and only if

$$|a - b| = |c - d|$$

Accordingly, we denote the orbits by E_1, E_2, $\cdots$, E_n, where

$$E_i = \{ab : |a - b| = i\}.$$

Then we distinguish three different types of edge orbits.

(i) E_n, the *short orbit*, is a one-factor, stabilized by α.

(ii) Each E_{2j+1}, $2j + 1 < n$, is an *odd* orbit. Each E_{2j+1} splits into two one-factors, $E_{2j+1} = F_{1j} \cup F_{2j}$ say, where

$$\begin{aligned}
F_{ij} &= \big\{(2i, 2i + 2j + 1) : 0 \le i < n\big\}, \\
F_{2j} &= \big\{(2i + 1, 2i + 2j + 2) : 0 \le i < n\big\}
\end{aligned}$$

These factors are preserved by α.

(iii) Each E_{2j}, $2j > n$, is an *even* orbit.

In a cyclic one-factorization, it must be possible to partition the union of all the even orbits, together with some or all of E_n and the odd orbits, into one-factors which are preserved by α.

We now examine the action of α on the set of one-factors in $\mathcal{F}$. Suppose the orbit of the factor F contains k elements, F, $F\alpha$, $F\alpha^2$, ..., $F\alpha^{k-1}$. Then α^k stabilizes F. It follows that k must be a factor of the order of α, which is $2n$.

Say F contains the edge xy. Then the factors $F\alpha$, $F\alpha^2$, $\cdots$ contain the edges $(x + 1, y + 1)$, $(x + 2, y + 2)$, $\cdots$ and all members of $E_{|x-y|}$. It follows that the union of all members of a factor orbit is a union of complete edge orbits. Each factor contains n edges, and each edge orbit except E_n contains $2n$ edges. So there will be exactly one factor orbit containing an odd number of factors, namely the orbit which contains E_n. Every other factor orbit has even order.

Suppose xy is an even edge in the factor F, and suppose k is the number of factors in the orbit of F. Then $(x + ik, y + ik)$ must also be an edge of F. There will be $\frac{2n}{k}$ members of $E_{|x-y|}$ in F. These edges contain between them $\frac{2n}{k}$ vertices of K_{2n}. Now let us further suppose k is even. Then either x is even, y is even, and all the vertices in the $\frac{2n}{k}$ members of $E_{|x-y|} \cap F$ are even; or else they are all odd.

Half the vertices of K_{2n} are even and half are odd. Every odd edge in F contains one even and one odd vertex. So the collection of all even edges in F must also contain equally many odd and even vertices. Say there are t even edges in F. They contain t even vertices. As we have pointed out, these vertices can be partitioned into $\frac{tk}{4n}$ sets of size $\frac{4n}{k}$, each being the set of vertices in edges in some even edge orbit. The t odd vertices in the even edges can be partitioned similarly, into $\frac{tk}{4n}$ further sets. It follows that there are an even number of even edge orbits represented in F — twice the integer $\frac{tk}{4n}$. We have proven

Lemma 15.1 [85] *If a factor orbit of $\mathcal{F}$ under α is of even order then it contains an even number of even edge orbits.* □

Corollary 15.1.1 [85] K_{2n} *has no cyclic one-factorization if $n = 2^t$, $t \geq 2$.*

Proof. If $n = 2^t$ then K_{2n} has $2^{t-1}-1$ even edge orbits, which is an odd number whenever $t \geq 2$. However, the length of any factor orbit must be a divisor of $2n$, the order of α, and all such divisors are even except for the divisor 1. A factor orbit of length 1 contains no even edge orbit (the only possible orbit of length 1 consists of the short orbit alone), so all orbits containing even edge orbits are of even order, and by Lemma 15.1 they will contain between them an even number of even orbits. But they must contain all even edge orbits — a contradiction. □

Theorem 15.2 [85] *There is a cyclic one-factorization of K_{2n} for every positive integer n not equal to 2^t, $t \geq 2$.*

Proof. The construction depends on the highest power of 2 to divide n.

(i) Suppose n is odd. Write S for the set of edges

$$(0,n), (1,-1), (2,-2), \cdots, (\lfloor \tfrac{n}{2} \rfloor, -\lfloor \tfrac{n}{2} \rfloor)$$

(reduced modulo 2^n) and $S + n$ for

$$\{(x+n, y+n) : (x,y) \in S\}.$$

Then $F = S \cup (S + n)$ is a one-factor of K_{2n}. (The edge $(0, n)$ is common to S and $S + n$, but is of course only counted once.) F, $F\alpha$, $F\alpha^2$, $\cdots F\alpha^{n-1}$ are distinct one-factors of K_{2n} and their union comprises E_n and all the even edge orbits. To form a one-factorization, append the factors F_{1j} and F_{2j} for every $j < \frac{n-1}{2}$. This factorization is clearly cyclic.

(ii) Suppose $n = 2m$, m odd. Write S for the set of edges

$$(0,m), \{1,-1\}, (2,-2), \cdots, (m-1, 1-m),$$

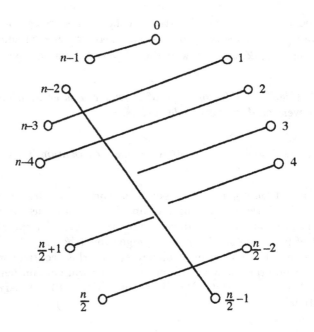

Figure 15.1   Edges in the set S.

and again write $S + n$ for $\{(x+n, y+n) : (x,y) \in S\}$.

Then the factors $F = S \cup (S+n)$, $F\alpha$, $\cdots$, $F\alpha^{n-1}$ contain all even edge orbits and the odd edge orbits E_m. Append the short edge orbit E_n and the one-factors from the remaining odd edge orbits to construct a cyclic one-factorization.

(iii) Suppose $n = 2^t m$, where $t \geq 2$ and $m \geq 3$. Write S for the set of edges

$$(0, n-1), (\tfrac{n}{2} - 1, n-2), (1, n-3), (2, n-4), \cdots, (\tfrac{n}{2} - 2, \tfrac{n}{2}),$$

as shown in Figure 15.1, and write

$$F = S \cup (S + 2^t m) = S \cup (S + n).$$

Then the factors

$$F, F\alpha, F\alpha^2, \cdots, F\alpha^{n-1}$$

contain all members of all even edge orbits except E_{n-2}, and also the even edge orbits E_{n-1} and $E_{\frac{n}{2}-1}$.

In addition to these n factors we take m more factors $G, G_1, \cdots, G_{m-1}$, where T is the set of edges

$$(0, n), (1, n-1), (2, m-2), (3, m-3), \cdots, (\lfloor \tfrac{m}{2} \rfloor, \lfloor \tfrac{m}{2} \rfloor + 1),$$

$$G = T \cup (T+m) \cup \ldots \cup (T + (2^t - 1)m),$$

and

$$G_1 = G^\alpha, G_2 = G_1 \alpha, \cdots.$$

These new factors cover E_{n-2}, the short edge orbit E_n and the odd edge orbits of lengths $1, 3, 5, \cdots, m-4$ (none of which have been covered by the F_i, since $m - 4 < \tfrac{n}{2} - 1$).

The remaining edges are all odd, and can be covered by factors comprising certain odd edge orbits. □

Hartman and Rosa [85] have enumerated the cyclic one-factorizations of K_{2n} for $2n \leq 16$. Of course there is none for $2n = 8$ or 16. They found that the examples for $2n = 4, 6, 10$ are unique up to isomorphism, and there are six isomorphism classes for $2n = 12$, and two for $2n = 14$.

Exercises 15

15.1 A one-factorization is called *transitive* if its automorphism group is transitive on the vertices.

   (i) Prove that the geometric factorization $\mathcal{GG}_{2n}$ is transitive.

   (ii) Hence show that there is a transitive one-factorization of K_{2n} for every n.

15.2 Suppose $\mathcal{F}$ is a one-factorization of K_{2n} based on $\mathbb{Z}_{2n}$, and α is the map $\alpha(i) = i + 1 \pmod{2n}$. We say a factor F in $\mathcal{F}$ is *stable* if $F\alpha = F$.

   (i) Prove that a cyclic one-factorization can have at most one stable factor.

   (ii) Prove that if $2n \equiv 0 \pmod 8$ then a cyclic one-factorization of K_{2n} cannot have a stable factor.

   (iii) Prove that if n is odd and if the cyclic one-factorization $\mathcal{F}$ of K_{2n} has a stable factor F, then α^n does not map any one-factor other than F in $\mathcal{F}$ onto itself.

(iv) Prove that there is a cyclic one-factorization $\mathcal{F}$ of K_{2n} with a stable factor, such that α^n maps every factor in $\mathcal{F}$ into itself, if and only if either $2n = 2$ or $2n \equiv 4 \pmod{8}$. [104]

16

PERFECT FACTORIZATIONS

Recall that a one-factorization is called *perfect* if it contains no two-division; in other words, the union of any two factors is a Hamilton cycle. Perfect one-factorizations exist for many orders, and we know of no order n (greater than 1) for which no perfect one-factorization of K_{2n} exists, but the existence question is not yet settled.

Theorem 16.1 *If p is an odd prime, then K_{p+1} and K_{2p} have perfect factorizations.*

Proof. The one-factorization $\mathcal{G}K_{p+1}$ of K_{p+1} from the patterned starter is perfect when p is prime, from Corollary 11.1.1. So we need only consider K_{2p}. We prove that $\mathcal{G}\mathcal{A}_{2p}$ is perfect when p is an odd prime. We adopt the notation of the "n odd" case in the definition of $\mathcal{G}\mathcal{A}_{2n}$ in Chapter 3, with n replaced by p. It is necessary to show that $J_i \cup J_k$ is a Hamilton cycle whenever $i \neq k$.

First suppose $1 \leq i \leq p$ and $1 \leq k \leq p$. We know that $P_{1,i} \cup P_{1,k}$ and $P_{2,i} \cup P_{2,k}$ are $(p+1)$-cycles. Figure 16.1 shows how $J_i \cup J_k$ is constructed from $(P_{1,i} \cup P_{1,k}) \cup (P_{2,i} \cup P_{2,k})$; it is clearly Hamiltonian.

Next, suppose both i and k are greater than p: say $i = p + e$ and $k = p + f$. Then j_i has edges $((p+x)_1, (p+x+e)_2)$ and J_k has edges $((p+x)_1, (p+x+f)_2)$ for all x. So the sequence of vertices in $J_i \cup J_k$ is

$$(p+1)_1, (p+1+e)_2, (p+1+e-f)_1, (p+1+2e-f)_2, (p+1+2e-2f)_1, \cdots$$

where the vertex label is reduced modulo p when necessary so that it lies in the range from $p + 1$ to $2p$. This sequence repeats itself after $2h$ steps only if

125

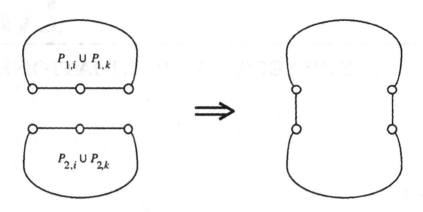

Figure 16.1   Showing that $\mathcal{G}\mathcal{A}_{2p}$ is perfect.

p divides $h(e - f)$; since e and f are different and p is prime, p must divide h, so $J_i \cup J_k$ is Hamiltonian.

Finally there is the case where $i \leq p$ and $j > p$. This is another simple piece of arithmetic, and is left as an exercise.   □

Theorem 16.1 is due to Anderson ([6, 10]). An alternative construction for a perfect factorization of K_{2p} was given by Nakamura [120], but Kobayashi [101] has shown that his factorizations are isomorphic to $\mathcal{G}\mathcal{A}_{2p}$ (see Exercises 16.7 and 16.8).

Apart from these two constructions, all other known perfect one-factorizations have been found by ad hoc methods, using starters. Anderson [7, 8, 12] found examples of orders 16, 28, 244 and 344. Orders 36 and 40 are due to Seah and Stinson [138, 139], and a different example of order 36 appears in [102]. Order 50 is constructed in [96]. Orders 1332 and 6860 are from [103], while orders 126, 170, 730, 1370, 1850, 2198 and 3126 come from [56]. Table 16.1 lists orders less than 100, and in each case gives a construction for a perfect one-factorization (K for $\mathcal{G}\mathcal{K}_{2n}$, A for $\mathcal{G}\mathcal{A}_{2n}$ if $\mathcal{G}\mathcal{K}_{2n}$ is not perfect, or a reference number for ad hoc constructions) or a ? if none is known.

A factorization is called *uniform* if $F_i \cup F_j$ is isomorphic to $F_k \cup F_\ell$ whenever $i \neq j$ and $k \neq \ell$. If it is known that a one-factorization is uniform then it is easy to check whether or not it is perfect: one only needs to test one pair of

n		n		n		n	
4:	K	28:	[8]	52:	?	76:	?
6:	K	30:	K	54:	K	78:	?
8:	K	32:	K	56:	?	80:	K
10:	A	34:	A	58:	A	82:	A
12:	K	36:	[138]	60:	K	84:	K
14:	K	38:	K	62:	K	86:	A
16:	[7]	40:	[139]	64:	?	88:	?
18:	K	42:	K	66:	?	90:	K
20:	K	44:	K	68:	K	92:	?
22:	A	46:	A	70:	?	94:	A
24:	K	48:	K	72:	K	96:	?
26:	A	50:	[96]	74:	K	98:	K

Table 16.1   Constructions for perfect one-factorizations.

factors. For this reason there have been a number of attempts to find uniform factorizations. The greatest success so far is:

Theorem 16.2 *If a is not a quadratic element in $GF(p^r)$, then the Mullin-Nemeth starter S_a generates a uniform one-factorization of K_{p^r+1}.*   □

The proof is left as an exercise.

Unfortunately no theoretical method has been found for predicting which Mullin-Nemeth starters generate perfect one-factorizations. The above Theorem has given a few of the sporadic values listed, but is most useful when $r = 1$, so that p is a prime. Although no new orders can arise, non-isomorphic perfect factorizations can be found; for example, Dinitz [49] finds three non-isomorphic one-factorizations of K_{20} and five of K_{24} (his examples are, furthermore, pairwise orthogonal).

There are only a few other results known about numbers of non-isomorphic perfect one-factorizations. In particular, complete searches have been done only as far as K_{14}: there is a unique perfect one-factorization (up to isomorphism) of K_4, K_6, K_8, and K_{10}, and there are precisely five isomorphism classes of perfect one-factorizations of K_{12} and 23 of K_{23}. (The results for K_{10}, K_{12} and K_{14} are from [73], [126] and [50] respectively.)

Why are perfect one-factorizations so hard to find? Most researchers would conjecture that they exist for all orders, but only two infinite classes are known and the sporadic results are very few. To investigate this situation we look at the automorphisms of perfect one-factorizations.

Lemma 16.3 [93] *Suppose $\mathcal{F}$ is a perfect one-factorization. The only automorphism of $\mathcal{F}$ which fixes two factors and one vertex is the identity.*

Proof. Suppose α is an automorphism which fixes factors F_1 and F_2 and vertex x. $F_1 \cup F_2$ is a cycle: say it is $x = x_1, x_2, \cdots, x_{2n}$, where $x = x_1 x_2$ is an edge of F_1. Then $x\alpha = x$ because α fixes x, $(xx_2)\alpha = xx_2$ because α fixes F_1, so $(xx_2)\alpha = xx_2$ and $x_2\alpha = x_2$. Since α fixes $F_2, (x_2x_3)\alpha = x_2x_3$ and $x_3\alpha = x_3$, and so on round the cycle; α is the identity.   □

Corollary 16.3.1 [93] *If $\mathcal{F}$ is a perfect one-factorization of K_{2n} then $|Aut(\mathcal{F})|$ divides $2n(2n-1)(2n-2)$.*

Proof. Consider the action of $Aut(\mathcal{F})$ on the set S of all ordered triples consisting of a vertex and two factors in $\mathcal{F}$. The only automorphism that fixes any member of S is the identity. So the orbits in S of $Aut(\mathcal{F})$ must all have the same number of elements as $Aut(\mathcal{F})$ itself. But the orbits partition S, so this common size must divide $|S|$, which equals $2n(2n-1)(2n-2)$.   □

These results are enough to suggest that standard methods are unlikely to produce further families of perfect factorizations. Most methods of construction involve subfactorizations or starters, and both these methods tend to give large groups. Ihrig [93, 94, 95] has carried out a detailed study of automorphisms of perfect one-factorizations, which gives more evidence for this suggestion. He proves the following improvement of Corollary 16.3.1:

Theorem 16.4 [95] *If $\mathcal{F}$ is a perfect one-factorization of K_{2n} then $|Aut(\mathcal{F})|$ divides one of $2n(2n-1), 2n(n-1)$ or $(2n-1)(2n-2)$.*   □

He also proved:

Theorem 16.5 [93] *The only perfect one-factorization of K_{2n} with a doubly-transitive automorphism group is $\mathcal{G}K_{2n}$ where $2n-1$ is prime.*   □

Theorem 16.6 [93] *If n is even, and $\mathcal{F}$ is a perfect one-factorization of K_{2n} with automorphism group of even order greater than $(2n-2)$, then $\mathcal{F}$ is $\mathcal{G}K_{2n}$ and $2n-1$ is prime.* □

Theorem 16.7 [95] *If $\mathcal{F}$ is a perfect one-factorization of K_{2n} which is not a member of the families $\mathcal{G}K$ or $\mathcal{G}A$, then $|Aut(\mathcal{F})|$ is at most $(2n-1)(n-1)$. If this maximum is attained then the automorphism group must act two-homogeneously on the one-factors, $2n-1$ must be a prime power, and $\mathcal{F}$ comes from a starter in an elementary abelian group.* □

The interested reader should consult the original papers [93, 95] and the survey [94].

Exercises 16

16.1 Prove that the one-factorization of K_{p+1} from the patterned starter in $\mathbb{Z}_p$ is perfect if and only if p is an odd prime.

16.2 If p is prime, $1 \le i \le p$ and $k > p$, prove that the union of factors J_i and J_k of $\mathcal{G}A_{2p}$ is a Hamilton cycle.

16.3 Construct two perfect factorizations of K_{14}, using the facts that $14 = 13+1$ and $14 = 2 \times 7$ respectively. Are the one-factorizations isomorphic?

16.4 Find two one-factors in $\mathcal{G}K_{10}$ whose union is not Hamiltonian.

16.5 Find two one-factors in $\mathcal{G}A_{18}$ whose union is not Hamiltonian.

16.6 Prove that the perfect one-factorization of K_8 is unique up to isomorphism.

16.7 Suppose p is an odd prime. If s is even, define

$$G_s = \{ij : i+j \equiv s, i \not\equiv j \,(\text{mod}\, 2p)\} \cup \{(\tfrac{s}{2}, p+\tfrac{s}{2})\}$$

and if s is odd, define

$$G_s = \{ij : i \text{ odd}, i - +j \equiv s\,(\text{mod}\, 2p)\}.$$

Prove that $G_0, G_1, \cdots, G_{p-1}, G_{p+1}, \cdots, G_{2p-1}$ form a one-factorization. [120]

16.8 Prove that the one-factorization in Exercise 16.7 is isomorphic to $\mathcal{G}\mathcal{A}_{2p}$. [101]

16.9 Prove that the one-factorization formed by developing the starter

$$
\begin{array}{cccccc}
(\infty, 0) & (1, 23) & (2, 16) & (3, 29) & (4, 21) & (5, 7) \\
(6, 13) & (8, 24) & (9, 34) & (10, 33) & (11, 17) & (12, 27) \\
(14, 15) & (18, 26) & (19, 22) & (20, 31) & (25, 30) & (28, 32)
\end{array}
$$

modulo 35 is perfect.

16.10 A *rotational* factorization of K_{2n} is a one-factorization consisting of the factor

$$(\infty_1, \infty_2), (0, n-1), (1, n), \cdots, (n-2, 2n-3)$$

and $2n - 2$ further factors formed by developing $(\bmod 2n - 2)$ some one-factor (called the *starter*) on the vertices $\infty_1, \infty_2, 0, 1, \cdots, 2n - 3$. (Both ∞_1 and ∞_2 are invariant under addition.) Prove that

$$
\begin{array}{cccccc}
(\infty_1, 0), & (\infty_2, 11), & (1, 2), & (3, 5), & (4, 24), & (6, 9), \\
(7, 22), & (8, 18), & (10, 17), & (12, 25), & (13, 21), & (14, 23), \\
(15, 31), & (16, 28), & (19, 30), & (20, 26), & (27, 32), & (29, 33)
\end{array}
$$

is a starter for a perfect rotational factorization of K_{36}. [102]

16.11 Prove that there exist non-isomorphic perfect factorizations of K_{36}.

16.12 Prove that geometric factorizations are uniform.

16.13 Prove Theorem 16.2.

16.14 Suppose $\mathcal{F}$ is a perfect one-factorization. Prove that the only automorphism of $\mathcal{F}$ which fixes three vertices is the identity. [94]

17

ONE-FACTORIZATIONS OF MULTIGRAPHS

The complete multigraph λK_v has v vertices and there are λ edges joining each pair of vertices. One-factors and one-factorizations of λK_v are defined in the obvious way; if v is odd there can be no one-factor, and when v is even, say $v = 2n$, one can produce a one-factorization of λK_{2n} by taking λ copies of each factor in a one-factorization of K_{2n}.

Given a one-factorization of λK_{2n}, it may be that there exists an integer λ_1 (smaller than λ) such that some $\lambda_1(2n-1)$ of the one-factors form a one-factorization of $\lambda_1 K_{2n}$. In that case the one-factorization of λK_{2n} is called *decomposable*; otherwise it is *indecomposable*. When $\lambda > 1$, the one-factorizations of λK_{2n} just exhibited are all decomposable. It is natural to ask for which values of λ and n do there exist indecomposable one-factorizations of λK_{2n}.

A one-factorization is called *simple* if it contains no repeated one-factor. There is no direct correspondence between simplicity and indecomposability. However simple one-factorizations are useful in the discussion of indecomposable one-factorizations.

We shall introduce λ-*starters*, analogous to starters. If G is a group of order 2n-1, then a λ-starter for G is a set of λ one-factors $F_1, F_2, \cdots, F_\lambda$ of the K_{2n} based on $G \cup \{\infty\}$ with the property that $F_1 \cup F_2 \cup \cdots \cup F_\lambda$ contains precisely λ edges xy such that $x - y = \pm g$, for every non-zero element g of G. Defining ∞ to be invariant under addition,

$$\mathcal{F} = \{F_i + g : 1 \leq i \leq \lambda, g \in G\}$$

is a one-factorization of λK_{2n}.

Theorem 17.1 [13] *If* $F = \{F_1, F_2, \cdots, F_\lambda\}$ *is a λ-starter in a group G of order $2n - 1$, and $\mathcal{F}$ is the one-factorization of λK_{2n} corresponding to F,*

(i) $\mathcal{F}$ *is simple if and only if $F_i = F_j + g$ never occurs when $g \in G$ and $i \neq j$;*

(ii) *If λ and $2n - 1$ are coprime and if there exist a non-zero element g in G and a factor F_i in F such that all λ edges xy for which $x - y = \pm g$ belong to F_i, then $\mathcal{F}$ is indecomposable.*   □

The proof is left as an exercise.

Theorem 17.2 [40] *There is a simple indecomposable one-factorization of $(n - 1)K_{2n}$ whenever $2n - 1$ is a prime.*

Proof. Suppose F is the one-factor

$$\infty 0, \ 12, \ 34, \ \cdots, \ (2n - 3)(2n - 2).$$

Let ξ be a generator of the multiplicative group of non-zero integers modulo $2n - 1$. Then define F_{ij} to be derived from F by multiplying each entry by ξ^i and then adding j: the edges in F_{ij} are

$$\infty j, \ (\xi^i + j)(2\xi^i + j), \ (3\xi^i + j)(4\xi^i + j), \ \cdots.$$

Then $\mathcal{F} = \{F_{ij} : 0 \leq i \leq n - 2, 0 \leq j \leq n - 2\}$ is a simple indecomposable one-factorization of $(n - 1)K_{2n}$: The result follows from Theorem 17.1.   □

Theorem 17.3 [13] *Given $\lambda > 2$, let p be the smallest prime which does not divide λ. Then there exists an indecomposable one-factorization of $\lambda K_{2(\lambda+p)}$.*

Proof. We construct an indecomposable one-factorization $\mathcal{F}$ of the $\lambda K_{2(\lambda+p)}$ based on two disjoint vertex-sets L and R, each containing $\lambda + p$ vertices. We label the vertices of L as x_L, where x ranges through $\mathbb{Z}_{\lambda+p}$, and similarly all members of R are labelled with a subscript R. All arithmetic is carried out modulo $\lambda + p$. $\mathcal{F}$ is based on the idea of a twin factorization.

For all i in $\mathbb{Z}_{\lambda+p}$ other than $i = 0$, define

$$G_i = \{(a_L, (a + i)_R) : a \in \mathbb{Z}_{\lambda+p}\}.$$

Each G_i is a one-factor of $\lambda K_{2(\lambda+p)}$. The first group of factors in $\mathcal{F}$ is the multiset consisting of G_1 taken $\lambda - p + 1$ times, G_{1-p} taken $\lambda - 1$ times, and each G_i, $3 \le i \le \lambda + p - 1$, taken $i + 1 - p$ times. Each edge incident with both of the sets L and R occurs in exactly λ of the factors, except for those of the form $(i_L, (i+1)_R)$, $(i_L(i-p+1)_R)$, $(i_L, (i+2)_R)$ and (i_L, i_R).

Next we describe factors $F_i, 0 \le i \le \lambda + p - 1$, which cover the remaining edges between L and R. F_0 consists of the $(i_L, (i+1)_R)$ for $0 \le i \le p - 2$, $((p-1)_L, 0_R)$, and all (j_L, j_R) for $p \le j \le \lambda + p - 1$; and for $1 \le i \le \lambda + p - 1$ define

$$F_i = \{((a+i)_L, (b+i)_R) : (a_L, b_R) \in F_0\}.$$

We include all these F_i, $0 \le i \le \lambda + p - 1$, in $\mathcal{F}$. By this stage $\mathcal{F}$ contains every edge from L to R exactly λ times, except that the edges $(i_L, (i+2)_R)$ are completely missing.

Now we construct the remaining factors. Notice that $\lambda + p$ is always odd: if λ is odd then $p = 2$ and if λ is even then p is odd. So both L and R contain an odd number of vertices. Select a near-one-factorization $\{H_1, H_2, \cdots, H_{\lambda+p}\}$ of the $K_{\lambda+p}$ based on L, and a near-one-factorization $\{K_1, K_2, \cdots, K_{\lambda+p}\}$ of the $K_{\lambda+p}$ based on R, with the property that for each i, i_R is the vertex missing from H_i and $(i+2)_R$ is the vertex missing from K_i. Then define

$$L_i = H_i \cup K_i \cup \{(i_L, (i+2)_R)\}.$$

$\mathcal{F}$ contains λ copies of each L_i, $0 \le i \le p + \lambda - 1$. This completes the description of $\mathcal{F}$, which is clearly a one-factorization of $\lambda K_{2(\lambda+p)}$.

We must prove that this factorization $\mathcal{F}$ is indecomposable. For this we will only need to use the factors F_i, $i = 0, \cdots, \lambda - p + 1$. Suppose that we can partition the factors of $\mathcal{F}$ into two sets $\mathcal{F}_1$ and $\mathcal{F}_2$ such that each edge of the underlying complete graph appears λ_i times in $\mathcal{F}_i$, $i = 1, 2$. Without loss of generality, suppose that F_0 lies in $\mathcal{F}_1$. This factor contains the edge (p_L, p_R) but does not contain $((p-1)_L, (p-1)_R)$. Because these edges are to occur in $\mathcal{F}_1$ with the same frequency, $\mathcal{F}_1$ must have a factor which does not contain (p_L, p_R) but does contain $((p-1)_L, (p-1)_R)$. But the only such factor is F_p. Thus $F_0 \in \mathcal{F}_1$ implies that $F_p \in \mathcal{F}_1$. Applying this argument again to F_p we prove that F_{2p} is in $\mathcal{F}_1$, and so on. Eventually we see that $F_i \in \mathcal{F}_1$ for $i = p, 2p, \cdots, (\lambda + p)p$. But p and $\lambda + p$ have greatest common divisor 1, so we have shown that $F_0, F_1, \cdots, F_{\lambda+p-1}$ all belong to $\mathcal{F}_1$. These factors cover the edge $(0_L, 0_R)$ λ times. It follows that $\mathcal{F}$ is indecomposable. $\square$

The next three Theorems provide recursive constructions: given an indecomposable one-factorization of some λK_{2n}, we build another. It is convenient to set up some notation. We write N_k for the set of the first k positive integers. If $\mathcal{F}$ is a one-factorization of the K_{2n} based on vertex-set N_{2n}, and U is any ordered $2n$-set, then $\mathcal{F}(U)$ is constructed by replacing i by the i-th member of U in every factor of $\mathcal{F}$, for every i. The factor derived from the factor F of $\mathcal{F}$ is denoted $F(U)$. If $\mathcal{L}$ is a one-factorization of the $K_{n,n}$ based on the two vertex-sets N_n and $N_{2n}\backslash N_n$, and U and V are ordered n-sets, then $\mathcal{L}(U,V)$ is the one-factorization formed from $\mathcal{L}$ by the substitutions

$$(1,2,\cdots,n) \quad \mapsto \quad U$$
$$(n+1,n+2,\cdots,2n) \quad \mapsto \quad V.$$

We say a one-factorization $\mathcal{F}$ of K_{2n} is *standardized* if the K_{2n} is based on N_{2n} and the i-th factor contains $(i,2n)$. A one-factorization of $K_{n,n}$ is standardized if the vertex-sets are N_n and $N_{2n}\backslash N_n$ and the first factor is

$$\{(1,n+1),(2,n+2),\cdots,(n,2n)\}.$$

The first two constructions use the idea of a twin factorization again.

Theorem 17.4 [14] *Suppose there exists an indecomposable one-factorization of λK_{2n} for some $\lambda > 1$. Then there exists an indecomposable one-factorization of λK_{4n} which is not simple.*

Proof. Suppose $\mathcal{F} = \{F_1, F_2, \cdots, F_{\lambda(2n-1)}\}$ is an indecomposable one-factorization of the λK_{2n} based on N_{2n}. Select two ordered $2n$-sets U and V, and a standardized one-factorization $\mathcal{L}$ of $K_{2n,2n}$. Then the factors in the one-factorization $\mathcal{L}(U,V)$, taken λ times each, together with the $\lambda(2n-1)$ factors $F_i(U) \cup F_i(V), 1 \leq i \leq \lambda(2n-1)$, form a non-simple one-factorization of λK_{4n}.

Suppose this factorization were decomposable: say $\{H_1, H_2, \cdots, H_s\}$ were a one-factorization of μK_{4n} where the H_i are among the factors listed. Write H_i' for the intersection of H_i with the K_{2n} based on U. Then $\{H_1', H_2', \cdots, H_s'\}$ is a one-factorization of the μK_{2n} based on U, and $\mathcal{F}(U)$ is indecomposable — a contradiction.   □

Theorem 17.5 [14] *Suppose there exists an indecomposable one-factorization of λK_{2n}, for some $\lambda > 1$. Then there exists an indecomposable one-factorization of λK_{4n-2} which is not simple.*   □

The proof is an easy generalization of the proof of Theorem 17.4, and is left as an exercise.

For the next result we use the fact that the graph $G(n, w)$, which was defined to have the integers modulo $2w$ as its vertices and edges xy whenever

$$w - n < x - y < w + n,$$

has a one-factorization whenever $n < w$. (See Theorem 3.3.)

A non-empty set of edges of a one-factor F is called a *subfactor* of F. A one-factorization $\mathcal{F}$ of the λK_{2n} based on V is said to be *contained* in a one-factorization $\mathcal{G}$ of the λK_{2s} based on W if $V \subseteq W$ and if for each factor F in $\mathcal{F}$ there is a one-factor G in $\mathcal{G}$ such that F is a subfactor of G. This is also expressed by saying that $\mathcal{F}$ is *embedded in* $\mathcal{G}$.

Theorem 17.6 [40] *Any indecomposable one-factorization of λK_{2n} can be embedded in a simple indecomposable one-factorization of λK_{2s} provided $\lambda \leq 2n - 1$ and $s \geq 2n$.*

Proof. Suppose $\mathcal{F} = \{F_{ij} : 1 \leq i \leq 2n - 1, 1 \leq j \leq \lambda\}$ is a one-factorization of the K_{2n} with vertex-set $V = \{v_1, v_2, \cdots, v_{2n}\}$. Since there are at most λ copies of any given factor, and $\lambda \leq 2n - 1$, we can assume that $F_{ij} \neq F_{im}$ when $j \neq m$. We construct a one-factorization of the K_{2s} with vertex-set $V \cup \mathbb{Z}_{2w}$, where $w \geq n$ and $\mathbb{Z}_{2w}$ is disjoint from V.

First take $H_1, H_2, \cdots, H_{2n-1}$ to be the factors in a one-factorization of $G(n, w)$. Write

$$K_{ij} = F_{ij} \cup H_i.$$

Then

$$\mathcal{K} = \{K_{ij} : 1 \leq i \leq 2n - 1, 1 \leq j \leq \lambda\}$$

is a set of one-factors of the K_{2s}. Moreover they are all different: if K_{ij} equals $K_{\ell m}$ then it must be that $H_i = H_\ell$ and $F_{ij} = F_{\ell m}$; the former implies that $i = \ell$, but $F_{ij} = F_{im}$ implies $i = m$.

We next need a sequence $(\{a_1, b_1\}, \{a_2, b_2\}, \cdots, \{a_{w-n}, b_{w-n}\})$ of disjoint pairs of non-zero elements of $\mathbb{Z}_{2w}$, such that $|a_r - b_r| = r$ for each r. This is easy: the pairs

$$\{1, w\}, \{2, w - 1\}, \cdots, \{\lfloor \tfrac{w}{2} \rfloor, \lfloor \tfrac{w+3}{2} \rfloor\}, \{w + 1, 2w - 1\},$$
$$\{w + 2, 2w - 2\}, \cdots, \{\lfloor \tfrac{3w-1}{2} \rfloor, \lfloor \tfrac{3w+2}{2} \rfloor\}$$

have differences $1, 2, \cdots, w - 1$ once each; select the pairs with differences $1, 2, \cdots, w - n$ and label them appropriately. Then write Y for the set of remaining elements of $\mathbb{Z}_{2w}$: that is,

$$Y = \mathbb{Z}_{2w} \backslash \bigcup_{r=1}^{w-n} \{a_r, b_r\},$$

and label the elements of Y as $\{y_1, y_2, \cdots, y_{2n}\}$. Now define

$$M_{ij} = \{\{v_t, y_{t+j} + i\} : 1 \leq t \leq 2n\} \cup \{\{a_r + i, b_r + i\} : 1 \leq r \leq w - n\}$$

(where the subscript on y is taken as an integer modulo $2n$, for the purposes of addition). Then if

$$\mathcal{M} = \{M_{ij} : i \in \mathbb{Z}_{2w}, 1 \leq j \leq \lambda\},$$

$\mathcal{M}$ is a set of $\frac{2w}{\lambda}$ distinct one-factors and $\mathcal{K} \cup \mathcal{M}$ is a simple one-factorization of λK_{2s} which contains the one-factorization $\mathcal{F}$.

Suppose $\mathcal{K} \cup \mathcal{M}$ is decomposable, and suppose certain of the factors constitute a factorization of μK_{2s}. Then the intersections of those factors with the K_{2n} based on V will be one-factors of the K_{2n} and will form a one-factorization of μK_{2n} as part of $\mathcal{F}$. Thus, if $\mathcal{F}$ is indecomposable, so is $\mathcal{K} \cup \mathcal{M}$.   □

Combining Theorems 17.3 and 17.6, one obtains:

Corollary 17.6.1 λK_{2n} *has a simple indecomposable one-factorization whenever* $2n \geq 4(\lambda + p), \lambda > 2$ *and* p *is the smallest prime not dividing* λ.

In addition to those already given, a number of other indecomposable factorizations have been constructed, mostly by computer. (It is easy to construct λ-starters, using a program similar to the strong starter algorithm.) The following result combines constructions from [40] and [13]; the one-factorization of $5K_8$ is previously unpublished.

Theorem 17.7 *There exist simple indecomposable one-factorizations of* $2K_8$, $4K_8$, $2K_{10}$, $4K_{10}$, $5K_{10}$, $3K_{12}$, $4K_{12}$, $6K_{12}$, $8K_{12}$, $9K_{12}$, $3K_{14}$, $8K_{14}$, $9K_{14}$, $10K_{14}$, $5K_{16}$, $7K_{16}$, $8K_{16}$, $9K_{16}$, $10K_{16}$, $7K_{18}$, $8K_{18}$, $9K_{18}$, $10K_{18}$, $7K_{20}$, $8K_{20}$, $9K_{20}$, $10K_{20}$, $7K_{22}$, $10K_{22}$, $7K_{24}$, $10K_{24}$, $7K_{26}$ *and* $7K_{28}$, *and indecomposable (but not simple) one-factorizations of* $5K_8$, $6K_8$ *and* $12K_{16}$.

Proof. The tables in Appendices B and C give a λ-starter for each factorization. $\qquad\square$

Observe that indecomposable (but not simple) one-factorizations of $2K_{10}$, $3K_{14}$ and $4K_{14}$ are given by Theorem 17.5.

It is easy to show that no indecomposable one-factorizations of $3K_6$ or $4K_6$ exist; we shall in fact prove a more general result in Theorem 17.11. Using this and the preceding results, we have:

Theorem 17.8 *An indecomposable one-factorization of λK_{2n} exists as follows:*

$$
\begin{aligned}
\lambda &= 2: &&\text{if and only if } 2n \geq 6; \\
\lambda &= 3: &&\text{if and only if } 2n \geq 8; \\
\lambda &= 4: &&\text{if and only if } 2n \geq 8; \\
\lambda &= 5: &&\text{if } 2n \geq 8; \\
\lambda &= 6: &&\text{if } 2n = 8 \text{ or } 2n \geq 12; \\
\lambda &= 7: &&\text{if } 2n \geq 16; \\
\lambda &= 8: &&\text{if } 2n \geq 12; \\
\lambda &= 9: &&\text{if } 2n \geq 12; \\
\lambda &= 10: &&\text{if } 2n \geq 14; \\
\lambda &= 11: &&\text{if } 2n = 24 \text{ or } 2n \geq 46; \\
\lambda &= 12: &&\text{if } 2n = 16 \text{ or } 2n \geq 30.
\end{aligned}
$$

$\qquad\square$

Since there are exactly $1 \cdot 3 \cdots (2n-1)$ one-factors of K_{2n}, the largest λ such that λK_{2n} has a simple factorization is

$$\lambda = 1 \cdot 3 \cdots (2n-3),$$

and for a simple indecomposable factorization we must have

$$\lambda < 1 \cdot 3 \cdots (2n-3).$$

However, this bound does not apply to indecomposable factorizations when simplicity is not required. We shall now derive a bound (which is probably very coarse) in that more general case.

By an *exact cover of depth d* on a set S we mean a collection of subsets of S, called blocks, such that each member of S belongs to exactly d blocks. (Repeated blocks are allowed.) If all the blocks are k-sets, the exact cover is called *regular of degree k*. An exact cover in S is *decomposable* if some proper subcollection of its blocks forms an exact cover on S. It is known (see [76]) that every sufficiently deep exact cover is decomposable: given s, there exists a positive integer $D[s]$ such that any exact cover of depth greater than $D[s]$ on an s-set is decomposable. It follows that there is also a maximum depth for a regular exact cover of degree k on an s-set: we denote it $D[s, k]$.

Lemma 17.9 [1] *Whenever $s \geq k \geq 1$,*

$$D[s, k] < s^s \cdot \binom{sk + s + 1}{s}.$$

$\square$

Theorem 17.10 *If there is an indecomposable factorization of λK_{2n}, then*

$$\lambda < [n(2n - 1)]^{n(2n-1)} \binom{2n^3 + n^2 - n + 1}{2n^2 - n}$$

Proof. Suppose there is an indecomposable factorization $\mathcal{F}$ of λK_{2n}. Denote by S the set of all edges of K_{2n}: S is a set of size $n(2n - 1)$. The factors in $\mathcal{F}$, interpreted as subsets of S, form an n-regular exact cover of depth λ on S. So

$$\lambda \leq D[n(2n - 1), n],$$

giving the result.

$\square$

As we said, this bound is very coarse. For example, in the case of λK_6, it is a little larger than 3×10^{31}. But we shall in fact show that no indecomposable one-factorization of λK_6 can exist for $\lambda \geq 3$. We assume that there is an indecomposable one-factorization $\mathcal{F}$ of λK_6 for some $\lambda \geq 3$ and derive a contradiction.

For notational convenience we assume K_6 to have vertices 0, 1, 2, 3, 4, 5. Since the fifteen one-factors of K_6 form a one-factorization of $3K_6$, not all of them can appear in $\mathcal{F}$: say $\{01, 23, 45\}$ is not represented. We denote the other possible one-factors as follows:

$$
\begin{array}{ll}
A = \{01, 24, 35\} & H = \{03, 15, 24\} \\
B = \{01, 25, 34\} & I = \{04, 12, 35\} \\
C = \{02, 13, 45\} & J = \{04, 13, 25\}
\end{array}
$$

$$D = \{02, 14, 35\} \qquad\qquad K = \{04, 15, 23\}$$
$$E = \{02, 15, 34\} \qquad\qquad L = \{05, 12, 34\}$$
$$F = \{03, 12, 45\} \qquad\qquad M = \{05, 13, 24\}$$
$$G = \{03, 14, 25\} \qquad\qquad N = \{05, 14, 23\}$$

Denote the number of occurrences of A in $\mathcal{F}$ as a, and so on.

Since edge 01 must appear in λ factors, we have

$$a + b = \lambda. \tag{17.1}$$

One could derive fourteen more equations in this way. In particular, considering 23, 02, 14 and 34 we get

$$k + n = \lambda \tag{17.2}$$
$$c + d + e = \lambda \tag{17.3}$$
$$d + g + n = \lambda \tag{17.4}$$
$$b +_e +\ell = \lambda \tag{17.5}$$

and $(17.1) + (17.2) + (17.3) - (17.4) - (17.5)$ is

$$a + c - g + k - \ell = \lambda \tag{17.6}$$

We can assume that $a \geq \frac{1}{2}\lambda$ and $k \geq \frac{1}{2}\lambda$: if $a < \frac{1}{2}\lambda$ and $k < \frac{1}{2}\lambda$, then carry out the permutation $(01)(45)$ on all members of $\mathcal{F}$ — it exchanges A with B and K with N and leaves $\{01, 23, 45\}$ unchanged; if $a < \frac{1}{2}\lambda$ and $k \geq \frac{1}{2}\lambda$ then (01) is the relevant permutation; if $a \geq \frac{1}{2}\lambda$ and $k < \frac{1}{2}\lambda$ then use (45).

The factors $\{A, C, G, K, L\}$ form a one-factorization of K_6, so a, c, g, k and ℓ cannot all be non-zero. The permutation $(01)(23)(45)$ exchanges G and L, and leaves A, C and $\{01, 23, 45\}$ unchanged; so without loss of generality we can assume $g \leq \ell$. Since a and k are positive, this means we can assume either c or g to be zero. But the equations derived from considering edges 03 and 45 are

$$f + g + h = \lambda \tag{17.7}$$
$$c + f = \lambda, \tag{17.8}$$

whence $h = c - g$ and $c \geq g$. So $g = 0$. Substituting this into (17.6) and recalling that $a \geq \frac{1}{2}\lambda$ and $k \geq \frac{1}{2}\lambda$ we obtain $c - \ell \leq 0$. Counting occurrences of 12 we see that

$$f + i + \ell = \lambda;$$

from (17.8) we get $i = c - \ell$, and as i cannot be negative we have also: $c = \ell$ and $a = k = \frac{1}{2}\lambda$. Considering (04) we get $i + j + k = \lambda$; this and (17.1) and

(17.2) now give $b = j = n = \frac{1}{2}\lambda$. Equation (17.5) tells us now that $e = a - c$, so $c \leq a \leq \frac{1}{2}\lambda$, and therefore from (17.8) f is non-zero. Since not all the members of the one-factorization $\{A, E, F, J, N\}$ can be represented, $e = 0$, whence c must equal $\frac{1}{2}\lambda$ also.

It is now easy to see that $e = g = i = m = 0$, and that the other ten factors each occur $\frac{1}{2}\lambda$ times. (The equations derived from edges 04 and 05 give the information about i and m.) If λ is odd, we have a contradiction. Otherwise we have $\frac{1}{2}\lambda$ duplicates of the one-factorization $\{A, B, C, D, F, H, J, K, L, N\}$ of $2K_6$, and $\mathcal{F}$ is decomposable. So we have proven the following Theorem.

Theorem 17.11 *There is no indecomposable one-factorization of λK_6 when $\lambda \geq 3$.*　　　　　　　　　　　　　　　　　　　　　　　□

Exercises 17

17.1 Prove that the factorization in Theorem 17.1 is indecomposable.

17.2 Prove Theorem 17.1.

17.3 Prove that there does not exist a simple or indecomposable one-factorization of λK_4 for any λ.

17.4 Find indecomposable one-factorizations of $2K_6$ and $3K_8$.

17.5 Prove Theorem 17.5.

17.6 If an indecomposable one-factorization $\mathcal{F}$ of λK_{2n} is embedded in a one-factorization $\mathcal{G}$ of λK_{2s}, prove that $\mathcal{G}$ is indecomposable.

17.7 Suppose $\mathcal{G}_1$, $\mathcal{G}_2$, $\mathcal{G}_3$ and $\mathcal{G}_4$ are indecomposable one-factorizations of $\lambda_1 K_{2n}$, $\lambda_2 K_{2n}$, $\lambda_3 K_{2n}$ and $\lambda_4 K_{2n}$ respectively, where $\lambda_1 + \lambda_2 = \lambda_3 + \lambda_4 = \lambda$, but $\lambda_3 \neq \lambda_1$ and $\lambda_3 \neq \lambda_2$. $\mathcal{H}$ is any twin factorization of λK_{4n} based on the vertex-sets U and V, using $\mathcal{G}_1 \cup \mathcal{G}_2$ on U and $\mathcal{G}_3 \cup \mathcal{G}_4$ on V. Prove that $\mathcal{H}$ is indecomposable.

18

MAXIMAL SETS OF FACTORS

One-factors are called *compatible* if they have no common edge; a set of one-factors of G is called *maximal* if the factors are compatible but G contains no other one-factor compatible with them. A one-factorization is trivially a maximal set; a maximal set of fewer than d one-factors in a regular graph of degree d will be called *proper*. Theorem 8.3 essentially tells us that $K_{n,n}$ has no proper maximal set (see Theorem 18.13, below). It is natural to ask about the existence of proper maximal sets in K_{2n}.

If S is a set of compatible one-factors of K_{2n}, we define the *leave* of S to be the complement of the union of members of S in K_{2n}, and denote it $\overline{S}$. Thus S is maximal if and only if S is a set of edge-disjoint one-factors and $\overline{S}$ contains no one-factor.

It is clear that no maximal set of one-factors in K_{2n} can contain $2n-2$ members — if S has $2n - 2$ members then $\overline{S}$ is regular of degree 1, and is itself a one-factor. The upper bound of $2n - 3$ factors can always be attained provided $n \geq 3$ (see Theorem 18.2 below). We now obtain lower bounds.

Theorem 18.1 [42] *If n is odd, then a maximal set in K_{2n} contains at least n one-factors. If n is even, then a maximal set in K_{2n} contains at least $n+1$ one-factors.*

Proof. Suppose S is a set of k disjoint one-factors of K_{2n}, where $k < n$. Then $\overline{S}$ is regular of degree $2n - k - 1$, and $2n - k - 1 \geq n$. By Corollary 2.3.1, S has a Hamilton cycle, and since the number of vertices is even that cycle decomposes into two one-factors, whence S is not maximal. So any maximal set in K_{2n} contains at least n one-factors.

141

It remains to show that K_{2n} contains no maximal set of n one-factors when n is even. But if S were such a set, $\overline{S}$ would be a regular graph of degree $n - 1$, on 2n vertices, n even, with no one-factor. This is impossible by Theorem 6.3.
□

Theorem 18.2 [42] *If e is odd and $e \leq n$ then K_{2n} contains a maximal set of $2n - e$ one-factors.*

Proof. We construct a set S of $2n - e$ compatible one-factors of K_{2n} for which $\overline{S}$ is $K_e \cup G$, G being some graph which is disjoint from the copy of K_e. Since $\overline{S}$ has odd components, it has no one-factor.

We partition the vertices of K_{2n} into two sets: V_1 is a set with e elements $\widehat{1}, \widehat{2}, \cdots, \widehat{e}$ and V_2 is a set with $2n - e$ elements $1, 2, \cdots, 2n - e$. Then we define F_i to be the one-factor with edges

$$\widehat{1}, i+1-e \quad \widehat{2}, i+2-e \quad \cdots \quad \widehat{e}, i \quad i+1, i-e \quad i+2, i-e-1$$
$$\cdots \quad i+n-e, i+n-e+1$$

where members of V_2 are treated as elements of $\mathbb{Z}_{2n-e}$ and reduced when necessary. (A copy of F_0 is shown in Figure 18.1. F_i is obtained by adding i to all the vertices in the right-hand column, and reducing modulo $2n - e$.) Then $F_0, F_1, \cdots, F_{2n-e-1}$ are clearly compatible and they together contain all edges joining V_1 to V_2 and no edges joining vertices of V_1. So these factors satisfy the requirements.
□

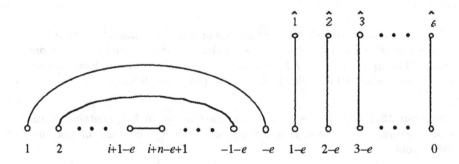

Figure 18.1 A factor in a maximal set.

The situation when e is even is quite different. To see this, it is easier to use the parameter $d = e - 1$, which is called the *deficiency* of the maximal set. The deficiency of S is then the degree of the regular graph $\overline{S}$. When e is even, the deficiency is odd. Since a regular graph of odd degree has no odd components, Theorem 7.1 applies:

Theorem 18.3 *If there is a maximal set of k one-factors in K_{2n}, and k is even, then*

$$k \geq \tfrac{4n+4}{3}.$$

Proof. Suppose a maximal set S of k one-factors exists. Then $\overline{S}$ is a regular graph of degree $d = 2n - k - 1$. If k is even, then $2n - k - 1$ is odd, and $\overline{S}$ has no odd component. By Theorem 7.1,

$$2n \geq 3(2n - k - 1) + 7,$$
$$2n \geq 6n - 3k + 4,$$

and therefore $k \geq \tfrac{4n+4}{3}$. $\qquad\qquad\square$

Our investigation of maximal sets of odd deficiency follows [130]. It is convenient to use the term *d-set* for a maximal set of one-factors of deficiency d.

Lemma 18.4 *Suppose there is a d-set in K_{2m}. Then there is a d-set in K_{2n} whenever $n \geq 2m$.*

Proof. From Theorem 14.2, there exists a one-factorization $\mathcal{F}$ of K_{2n} which contains a subfactorization of K_{2m}. Suppose the factors in $\mathcal{F}$ are $F_1, F_2, \cdots,$ F_{2n-1}, and say $F_1 \cap K, F_2 \cap K, \cdots, F_{2m-1} \cap K$ form a one-factorization of K, where K is a copy of K_{2m}. Select a d-set $G_1, G_2, \cdots, G_{2m-1}$ of one-factors of K, and for each i, $1 \leq i \leq 2m - 1 - d$, define

$$H_i = (F_i \backslash K) \cup G_i.$$

Then $H_1, H_2, \cdots, H_{2m-1-d}, F_{2m}, F_{2m-1}, \cdots, F_{2n-1}$ form a d-set in K_{2n}. $\quad\square$

We shall use a family of graphs R_d in our discussion. For each d, R_d is a regular graph of degree $2d + 6$ on $3d + 7$ vertices. We shall prove that R_d has a one-factorization, but its complement $\overline{R_d}$ has no one-factor.

It is easiest to describe R_d as the union of three graphs A_{d1}, A_{d2}, A_{d3}, a vertex ∞, and some further edges. A_{di} has vertices 0_i, 1_i, $\cdots$, d_i and a_i, where 0, 1, $\cdots$, d are the elements of $\mathbb{Z}_{d+1}$ and a_i is left invariant by the operations of $\mathbb{Z}_{d+1}$. The edges in A_{d1} and A_{d2} consist of disjoint edges and one path of length 2, namely

$$0_i 1_i, 2_i 3_i, \cdots, (d-3)_i (d-2)_i, (d-1)_i a_i d_i$$

($i = 1$ or 2), while A_3 is the edge $0_3 1_3$ and the path

$$(\tfrac{d+1}{2})_3, (\tfrac{d-1}{2})_3, \cdots, 3_3, 2_3, a_3, d_3, (d-1)_3 \cdots (\tfrac{d+5}{2})_3, (\tfrac{d+3}{2})_3.$$

Vertex ∞ is joined to 0_3, 1_3, $(\tfrac{d+1}{2})_3$, $(\tfrac{d+3}{2})_3$ and to all vertices x_1 and x_2, $x \in \mathbb{Z}_{d+1}$ (but not to a_1 or a_2). Finally, every edge joining vertices of A_{di} and A_{dj}, $i \neq j$, is included.

As an example, R_5 is shown in Figure 18.2. (A bold zigzag means that all edges joining members of the indicated sets are included.)

Lemma 18.5 $\overline{R}_d$ has no one-factor.

Proof. Consider the removal of vertex ∞ from $\overline{R}_d$. The leaves three odd components, namely $\overline{A}_1$, $\overline{A}_2$ and $\overline{A}_3$. So the Lemma follows from Theorem 6.1. □

Lemma 18.6 R_d has a one-factorization for every odd $d \geq 3$.

Proof. In each case we give two seed factors F_1 and F_2, which are to be developed modulo $d + 1$, yielding $2d + 2$ factor F_1, $F_1 + 1$, $\cdots$, $F_2 + d$ will then give two Hamilton cycles which contain the remaining edges. When $d = 3$, we use

$$F_1 = 1_1 1_2, \ 0_2 1_3, \ 2_2 0_3, \ 0_1 2_3, \ \infty 2_i, \ a_1 3_2, \ a_2 3_3, \ a_3 3_1$$
$$F_2 = 0_1 1_2, \ 0_2 0_3, \ 1_1 2_3, \ 3_1 3_3, \ \infty 2_2, \ a_1 1_3, \ a_2 2_1, \ a_3 3_2$$

and the cycles

$$0_1 1_1 0_2 2_1 1_2 0_3 1_3 2_2 3_1 2_3 \infty 3_3 a_3 a_1 a_2 3_2 0_1,$$
$$0_1 2_2 a_2 a_3 2_3 3_2 1_1 0_3 \infty 1_3 2_1 a_1 3_1 1_2 0_2 3_3 0_1.$$

The seed factors and cycles for cases $d \equiv 1 \bmod 4$ and $d \equiv 3 \bmod 4$, $d \geq 7$ are given in Tables 18.1 and 18.2. □

From Lemmas 18.4 and 18.6 we have:

F_1:

$$1_1 1_2 \qquad\qquad 2_2 0_3 \qquad\qquad (c+4)_1 (c+2)_3$$
$$3_1 5_2 \qquad\qquad 4_2 1_3 \qquad\qquad (c+6)_1 (c+3)_3 \quad \infty(c+2)_1$$
$$5_1 9_2 \qquad\qquad 6_2 2_3 \qquad\qquad (c+8)_1 (c+4)_3 \quad a_1(2c-1)_2$$
$$\vdots \qquad\qquad \vdots \qquad\qquad \vdots \qquad\qquad a_2(c+1)_3$$
$$(2c-1)_1(2c-5)_2 \quad (2c)_2(c-1)_3 \quad c_1(2c+1)_3 \quad a_3(2c+1)_1$$
$$0_2 c_3$$

F_2:

$$1_1 2_2 \qquad\qquad (c+3)_2 0_3 \qquad 0_1 c_3$$
$$3_1 6_2 \qquad\qquad (c+5)_2 1_3 \qquad 2_1(c+1)_3 \qquad \infty(c+1)_2$$
$$5_1 10_2 \qquad\qquad (c+7)_2 2_3 \qquad 4_1(c+2)_3 \qquad a_1(2c+1)_3$$
$$\vdots \qquad\qquad \vdots \qquad\qquad \vdots \qquad\qquad a_2(2c+1)_1$$
$$(2c-1)_1(2c-2)_2 \quad (c-1)_2(c-1)_3 \quad (2c)_1(2c)_3 \quad a_3(2c)_2$$

Hamilton cycles:

$d = 5$ $0_1 1_1 0_2 2_1 1_2 3_1 2_2 1_3 0_3 \infty 3_3 2_3 3_2 4_1 a_1 5_1 4_2 a_2 a_3 5_3 4_3 5_2 0_1;$
$0_1 4_2 3_3 4_1 2_2 3_2 5_1 4_3 \infty 1_3 2_1 3_1 2_3 a_3 a_1 a_2 5_2 1_1 0_3 1_2 0_2 5_3 0_1.$

$d = 9$: $0_1 1_1 0_2 2_1 1_2 3_1 2_2 1_3 0_3 \infty 5_3 4_3 3_3 2_3 3_2 4_1 5_1 4_2 5_2 6_1 7_1 6_2 7_2 6_3 7_3 8_1$
$a_1 9_1 8_2 a_2 a_3 9_3 8_3 9_2 0_1;$
$0_1 8_2 7_3 8_3 9_1 7_2 8_1 6_2 5_3 6_1 4_2 3_3 4_1 2_2 3_2 5_1 4_3 5_2 7_1 6_3 \infty 1_3 2_1 3_1 2_3 a_3$
$a_1 a_2 9_2 1_1 0_3 1_2 0_2 9_3 0_1.$

$d \geq 13$ $0_1 1_1 0_2 2_1 1_2 3_1 2_2 1_3 0_3 \infty[(c+1)_3 c_3 \cdots 3_3 2_3] 3_2 [4_1 5_1 4_2 5_2 \cdots$
$(c+2)_1 (c+3)_1 (c+2)_2 (c+3)_2] (c+2)_3 (c+3)_3 [(c+4)_1$
$(c+5)_1 (c+4)_2 (c+5)_2 (c+4)_3 (c+5)_3 \cdots (2c-2)_1$
$(2c-1)_1 (2c-2)_2 (2c-1)_2 (2c-2)_3 (2c-1)_3] (2c)_1 a_1$
$(2c+1)_1 (2c)_2 a_2 a_3 (2c+1)_3 (2c)_3 (2c+1)_2 0_1;$
$0_1 (2c)_2 [(2c-1)_3 (2c)_3 (2c+1)_1 (2c-1)_2 (2c)_1 (2c-2)_2$
$(2c-3)_3 (2c-2)_3 (2c-1)_1 \cdots (c+3)_3 (c+4)_1 (c+3)_2 (c+4)_1$
$(c+2)_2][(c+1)_3 (c+2)_1 c_2 (c-1)_3 c_1 (c-2)_2 \cdots 3_3 4_1 2_2][3_2 5_1$
$4_3 7_1 6_3 \cdots (c+1)_2 (c+3)_1 (c+2)_3] \infty 1_3 2_1 3_1 2_3 a_3 a_1 a_2$
$(2c+1)_2 1_1 0_3 1_2 0_2 (2c+1)_3 0_1.$

Table 18.1   Constituents of the factorization of R_d, $d = 2c + 1 \equiv 1 \bmod 4$.

F_1:

$\infty(c+1)_1$	$a_3(c+2)_1$	$a_1(2c+1)_2$	$a_2(2c+1)_3$
$1_1 1_2$	$(c+4)_1 3_2$	$2_2 0_3$	$(c+3)_1(c+1)_3$
$3_1 5_2$	$(c+6)_1 7_2$	$4_2 1_3$	$(c+5)_1(c+2)_3$
$5_1 9_2$	$(c+8)_1 11_2$	$6_2 2_3$	$\vdots$
$\vdots$	$\vdots$	$\vdots$	$0_1 \frac{1}{2}(3c+1)_3$
$c_1(2c-1)_2$	$(2c+1)_1(2c-3)_2$	$(2c)_2(c-1)_3$	$\vdots$
		$0_2 c_3$	$(c-1)_1(2c)_3$

F_2:

$\infty(c+1)_2$	$a_1 c_3$	$a_2(c+1)_1$	$a_3(2c+1)_2$
$0_1 1_2$	$(c+3)_1 3_2$	$(c+3)_2 0_3$	$1_1(c+1)_3$
$2_1 5_2$	$(c+5)_1 7_2$	$(c+5)_2 1_3$	$3_1(c+2)_3$
$4_1 9_2$	$(c+7)_1 11_2$	$(c+7)_2 2_3$	$5_1(c+3)_3$
$\vdots$	$\vdots$	$\vdots$	$\vdots$
$(c-1)_1(2c-1)_2$	$(2c)_1(2c-3)_2$	$0_2 \frac{1}{2}(c-1)_3$	$(2c+1)_1(2c+1)_3$
		$\vdots$	
		$(c-1)_2(c-1)_3$	

Hamilton cycles:

$d = 7$ $0_1 1_1 0_2 2_1 1_2 3_1 2_2 1_3 0_3 \infty 4_3 3_3 2_3 3_2 4_1 5_1 4_2 5_2 6_1 a_1 7_1 6_2 5_3 6_3 7_3 a_3 a_2 7_2 0_1$;

$0_1 6_2 a_2 a_1 a_3 2_2 3_1 2_1 1_3 \infty 5_3 6_1 4_2 3_3 4_1 2_2 3_2 5_1 4_3 5_2 7_1 6_3 7_2 1_1 0_3 1_2 0_2 7_3 0_1$.

$d = 11$ $0_1 1_1 0_2 2_1 1_2 3_1 2_2 1_3 0_3 \infty 6_3 5_3 4_3 3_3 2_3 3_2 4_1 5_1 4_2 5_2 6_1 7_1 6_2 7_2 8_1 9_1$
$8_2 7_3 8_3 9_2 10_1 a_1 11_1 10_2 9_3 10_3 11_3 a_3 a_2 11_2 0_1$;

$0_1 10_2 a_2 a_1 a_3 2_3 3_1 2_1 1_3 \infty 7_3 8_1 6_2 5_3 6_1 4_2 3_3 4_1 2_2 3_2 5_1 4_3 5_2 7_1 6_3 7_2$
$9_1 8_3 9_3 10_1 8_2 9_2 11_1 10_3 11_2 11_3 0_3 1_2 0_2 11_3 0_1$.

$d \geq 15$ $0_1 1_1 0_2 2_1 1_2 3_1 2_2 1_3 0_3 \infty [(c+1)_3 c_3 \cdots 3_3 2_3] 3_2 [4_1 5_1 4_2 5_2 \cdots (c+1)_1$
$(c+2)_1(c+1)_2(c+2)_2](c+3)_1(c+4)_1(c+3)_2(c+2)_3$
$(c+3)_3(c+4)_2(c+5)_1[(c+6)_1(c+5)_2(2c+4)_3(c+5)_3$
$(c+6)_2(c+7)_1 \cdots (2c-1)_1(2c-2)_2(2c-3)_3(2c-2)_3(2c-1)_2$
$(2c)_1]a_1(2c+1)_1(2c)_2(2c-1)_3(2c)_3(2c+1)_3 a_3 a_2(2c+1)_2 0_1$;
$0_1(2c)_2 a_2 a_1 a_3 2_3 3_1 2_1 1_3 \infty [(c+2)_3(c+3)_1(c+1)_2 c_3(c+1)_1(c-1)_2$
$\cdots 3_3 4_1 2_2][3_2 5_1 4_3 5_2 7_1 6_3 \cdots (c+2)_2(c+4)_1(c+3)_2][(c+4)_3$
$(c+5)_1(c+3)_2(c+4)_2(c+6)_1(c+5)_3 \cdots (2c-1)_3(2c)_1$
$(2c-2)_2(2c-1)_2(2c+1)_1(2c)_3](2c+1)_2 1_1 0_3 1_2 0_2(2c+1)_3 0_1$.

Table 18.2   Constituents of the factorization of R_d, $d = 2c+1 \equiv 3 \bmod 4$.

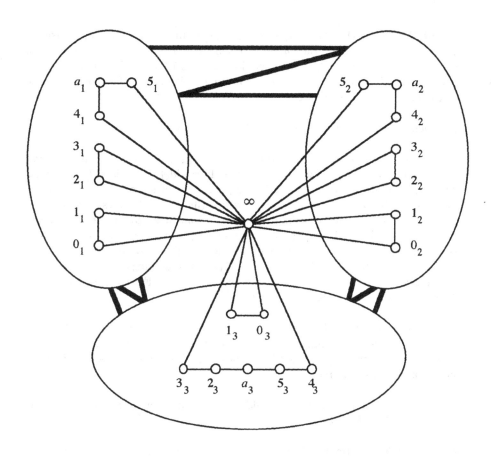

Figure 18.2 The graph R_5.

Theorem 18.7 *There is a d-set in K_{2n} when $2n = 3d + 7$ or $2n \geq 6d + 14$, for every odd $d \geq 3$.* □

It remains to handle the intermediate cases.

Theorem 18.8 *For each odd integer $d \geq 3$ and each even integer $2n$, $4d + 8 \leq 2n \leq 6d + 12$, there exists a d-set in K_{2n}.*

Proof. Let H be the graph $R_d + K_{2n-(3d+7)}$ (where $+$ is the join, as usual). Write $t = 2n - (3d + 7)$; then $d + 1 \le t \le 3d + 5$. Let $\{F_1, F_2, \cdots, F_{2d+6}\}$ be the one-factorization of R_d constructed in Lemma 18.6, and let $\{F'_1, F'_2, \cdots, F'_{t-(d+1)}\}$ be a collection of $t - (d+1)$ mutually disjoint one-factors in K_t. We construct the d-set as follows.

For each $i = 1, 2, \cdots, t - (d + 1)$ take the one-factors $E_i = F_i \cup F'_i$.

There remain in R_d the one-factors $F_{t-d}, F_{t-d+1}, \cdots, F_{2d+6}$ which among them contain $(3d + 7) \cdot \frac{1}{2}(2d + 6 - (t - d) + 1) = (3d + 7) \cdot \frac{1}{2}(3d + 7 - t)$ edges. Apply Theorem 8.12 with $c = 3d + 7$ to partition this set of edges into $3d + 7$ matchings $M_1, M_2, \cdots, M_{3d+7}$, each with $\frac{1}{2}(3d + 7 - t)$ edges. For each j, $j = 1, 2, \cdots, 3d + 7$, the matching M_j covers all but a set S_j of t vertices; since the union of all matchings M_j is a regular graph of valency $3d + 7 - t$ each vertex in R_d is contained in exactly t of the sets S_j. Now we apply Theorem 13.3 to construct a set $M'_1, M'_2, \cdots, M'_{3d+7}$ of matchings on H where each M'_j matches the vertices of S_j to the vertices of K_t. For each $j = 1, 2, \cdots, 3d + 7$ take the one- factor $E_j = M_j \cup M'_j$.

The set of one-factors $\{E_i : 1 \le i \le t - (d + 1)\} \cup \{E_j : 1 \le j \le 3d + 7\}$ then constitutes a maximal set (the complement of their union cannot contain a one-factor, since we have noted previously that $\overline{R_d}$ contains no one-factor), so it is a d-set as required. $\qquad \square$

For the cases where $3d + 9 \le 2n \le 4d + 6$ we define graphs R_d^w for every odd w, $3 \le w \le d$, as follows.

From R_d delete vertex 0_3. Next, consider the subgraph B_d of A_{d3} formed by deleting 0_3 and 1_3 (in other words, the subgraph of R_d spanned by the vertices not adjacent to or equal to 0_3). B_d is a path of length $d - 1$. Relabel its vertices so that the path becomes $0, 2, 4, \cdots, d-2$. Now for each i, $0 \le i \le w-2$, append to this subgraph the edges in $E(i)$, .

$$E(i) = \left\{ (i - j, i + j) : \tfrac{w+1}{2} \le j \le \tfrac{d-i}{2} \right\}.$$

(Notice that this process adds $\frac{1}{2}(w - 1)(d - w)$ new edges. When $w = d$ then each $E(i)$ is empty, so no new edges are added.)

For $0 \le i \le w - 2$ it is convenient to write $V(i)$ for the set of vertices covered by edges in $E(i)$ together with vertex i, that is

$$V(i) = \{i\} \cup \left\{ i - j, i + j : \tfrac{w+1}{2} \le j \le \tfrac{d-1}{2} \right\},$$

and also write

$$V(w-1) = \{w-1, w, \cdots, d-1\}.$$

Lemma 18.9 R_d^w has $3d + 7$ vertices and

$$\chi'(R_d^w) \le 2d + 5 + w.$$

Proof. First, $\chi'(R_d) = 2d + 6$, by Lemma 18.4, so certainly $R_d \backslash \{0_3\}$ requires at most $2d + 6$ colors. Each $E(i)$ is a set of independent edges, and can be colored in one color. So R_d^w requires at most $(2d + 6) + (w - 1) = 2d + 5 + w$ colors. □

Since R_d^w was constructed from R_d by deleting the $2d + 6$ edges incident with 0_3 and adding $\frac{1}{2}(w - 1)(d - w)$ new edges. So

$$
\begin{aligned}
e(R_d^w) &= \frac{1}{2}(3d + 7)(2d + 6) - (2d + 6) + \frac{1}{2}(w - 1)(d - w) \\
&= (2d + 5 + w)(\tfrac{3d+6-w}{2})
\end{aligned}
$$

and Theorem 8.12 applies to R_d^w. We can decompose the edge-set of R_d^w into $2d + 5 + w$ matchings, each with $\frac{3d+6-w}{2}$ edges. Select such a decomposition into matchings $M_1', M'_2, \cdots, M'_{2d+5+w}$. For each j, write S_j' for the set of vertices of R_d^w **not** incident with M_i'.

Now consider the sets $V(i)$. They form the rows in the following $w \times (d - w + 1)$ array of elements of $\mathbb{Z}_{d+1}$. Permute the elements of the first row and last column

0	$\frac{1}{2}(w + 1)$	$\frac{1}{2}(w + 3)$	$\cdots$	$d - \frac{1}{2}(w + 1)$
1	$\frac{1}{2}(w + 3)$	$\frac{1}{2}(w + 5)$	$\cdots$	$d - \frac{1}{2}(w - 1)$
$\vdots$	$\vdots$	$\vdots$		$\vdots$
$w - 2$	$\frac{1}{2}(3w - 3)$	$\frac{1}{2}(3w - 1)$	$\cdots$	$\frac{1}{2}(w - 5)$
$w - 1$	w	$w + 1$	$\cdots$	$d - 1.$

of this array by the mapping

$$k \mapsto k + \tfrac{1}{2}(w - 1) \bmod d.$$

The resulting array contains every symbol at most once in each column. Let S_i denote the set of vertices of R_d^w in column i of the new array.

Lemma 18.10 *Each vertex of R_d^w belongs to exactly w of the sets S_1, S_2, $\cdots$, S_{d-w+1}, S_1', S_2', $\cdots$, S_{2d+5+w}'.*

Proof. Consider a vertex x of R_d^w. If x is not a vertex of B_d then x has degree $2d + 5$ in R_d^w so it belongs to w of the sets S_j', and it is not a member of any of the S_i, since each S_i is a set of vertices of B_d. Now suppose x is a vertex of B_d. Then x has degree $2d + 6 + k$ in R_d^w, where k is the number of the $E(i)$ of which x is a vertex. Hence, the number of the matchings M'_j with which x is a vertex. Hence, the number of the matchings M'_j with which x is **not** incident is

$$2d + 5 - w - (2d + 6 + k)$$
$$= w - k - 1,$$

so x is in $w - k - 1$ of the sets S_j'. Now,

$$
\begin{aligned}
k &= \text{number of } E(i) \text{ incident with } x \\
&= (\text{number of } V(i) \text{ containing } x) - 1 \\
&= (\text{number of } S_i \text{ containing } x) - 1,
\end{aligned}
$$

so x is contained in $k + 1$ of the S_i, and in $w - k - 1 + k + 1 = w$ of the sets listed.   □

The sets S_i and S_j' are all w-sets, so they form a collection of $3d + 6$ w-sets which between them contain the $3d + 6$ vertices of R_d^w w times each.

Theorem 18.11 *For each odd integer $d \geq 3$ and each integer n such that $3d + 9 \leq 2n \leq 4d + 6$ there is a d-set in K_{2n}.*

Proof. We write $w = 2n - (3d + 6)$; then w is odd and $3 \leq w \leq d$. Consider the graph $J = R_d^w + \overline{K}_w$. From Theorem 13.3 and Lemma 18.10, the edges of J joining R_d^w to $\overline{K}_w$ can be partitioned into $3d + 6$ matchings M_1, M_2, $\cdots$, M_{3d+6} where M_j is a matching from S_j' to $\overline{K}_w$ whenever $j \leq 2d + 5 + w$. Since S_j' is the set of vertices not incident with the matching M'_j, $M_j \cup M'_j$ is a one-factor of J. We define a graph G to be the union of the $M_j \cup M'_j$, $1 \leq j \leq 2d+5+w$. Then $\overline{G}$ is a d-regular graph on $2n = 3d+6+w$ vertices. We shall show that $\overline{G}$ contains no one-factor, thus proving that the $M_j \cup M'_j$ form the required d-set.

G can be constructed from R_d by replacing the vertex 0_3 by w new vertices x_1, x_2, $\cdots$, x_w, replacing each edge $(v, 0_3)$ by w new edges (v, x_1), (v, x_2), $\cdots$, (v, x_w), and adding some edges to the subgraph spanned by the vertices of B_d

and $\{x_1, x_2, \cdots, x_w\}$. So removing the vertex ∞ from $\overline{H}$ will create at least two odd components, one on the vertices of A_{d1} and one on the vertices of A_{d2}. So, by Theorem 6.1, $\overline{H}$ has no one-factor. □

From Theorems 18.7, 18.8 and 18.11 we have:

Theorem 18.12 *There is a maximal set of k one-factors in K_{2n} whenever* $k \geq \frac{4n+4}{3}$. □

The situation for complete bipartite graphs is quite different from that for complete graphs. Theorem 8.3 tells us that any bipartite graph G satisfies $\chi'(G) = \Delta(G)$. In particular, if G is regular, G has a one-factorization. Since the leave in $K_{n,n}$ of any set of compatible one-factors of $K_{n,n}$ is a regular bipartite graph, it has a one-factorization; so *a fortiori* it has a one-factor:

Theorem 18.13 *There is no proper maximal set of one-factors in $K_{n,n}$.* □

This Theorem has a useful interpretation in design theory. We define a *Latin rectangle* of size $n \times r$, where $n > r$, to be an $n \times r$ array in which every column is a permutation of $\{1, 2, \cdots, n\}$ and no row contains a repeated symbol. When $n < r$, the definition is the same except that it is the rows which are complete permutations. Using the correspondence in Chapter 3 between Latin squares and one-factorizations of $K_{n,n}$, we see that a Latin rectangle corresponds to a set of r one-factors of $K_{n,n}$. So we have:

Corollary 18.13.1 *Every Latin rectangle can be extended to a Latin square.* □

Exercises 18

18.1 We say a set of one-factors in G is *Hamiltonian* if it contains two one-factors whose union is a Hamilton cycle in G.

    (i) Prove that the complete bipartite graph $K_{v,v}$ has a Hamiltonian one-factorization.

(ii) Suppose there exists a d-set in K_{2n}. Prove that there exists a Hamiltonian d-set in K_{4n}.

(iii) Suppose there exists a Hamiltonian d-set in K_{2n}. Prove that there exists a Hamiltonian d-set in K_{4n-2}. [34]

18.2 Figure 18.3 shows a set $S = \{S_1, S_2, \cdots, S_{14}\}$ of one-factors of the K_{18} with vertices 1, 2, 3, 4, 5, 6, 7, 8, 9, R, S, T, U, V, W, X, Y, Z, and its leave $\overline{S}$. Prove that S is a maximal set of deficiency d in K_{18}. [34]

S_1	=	12	3T	4Z	58	6R	7X	9U	SW	VY
S_2	=	13	2S	4W	5V	68	7R	9Z	TX	UY
S_3	=	14	29	38	5Y	6Z	7W	RX	SV	TU
S_4	=	15	2V	3U	4X	6W	78	9Y	RS	TZ
S_5	=	16	2R	3Y	47	59	8S	TV	UW	XZ
S_6	=	17	2X	35	46	8Y	9S	RZ	TW	UV
S_7	=	18	2Y	37	4T	5U	6S	9X	RV	WZ
S_8	=	19	26	3R	4U	5W	7Z	8T	SY	VX
S_9	=	1T	27	34	5Z	6Y	8X	9R	SU	VW
S_{10}	=	1U	2W	3X	49	5T	6V	7Y	8R	SZ
S_{11}	=	1V	23	48	57	69	RW	ST	UX	YZ
S_{12}	=	1W	25	39	4V	6X	7S	8Z	RU	TY
S_{13}	=	1X	24	3S	56	79	8U	RT	VZ	WY
S_{14}	=	1Y	28	36	4R	5S	7V	9T	UZ	WX

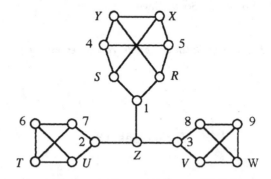

Figure 18.3 The set S and its leave $\overline{S}$.

19

THE ONE-FACTORIZATION CONJECTURE

The following conjecture has been made independently by a number of mathematicians (including the author, who was certainly not the first):

The One-Factorization Conjecture *If G is a regular graph of degree d on $2n$ vertices which satisfies*

$$d \geq n - 1 \quad \text{if } n \text{ is even}$$
$$d \geq n \qquad \text{if } n \text{ is odd}$$

then G has a one-factorization.

It follows from Theorem 18.2 that the bound in the conjecture is best-possible. The conjecture itself appears quite difficult. The case $d = 2n - 2$ is clearly true, since a regular graph of degree $2n - 2$ is just the complement of a one-factor in K_{2n}. The case of degree $2n - 3$ has been part of the combinatorial folklore. Case $d = 2n - 4$ was proven in [132] under the (not very restrictive) assumption that the complement $\overline{G}$ has a one-factorization. It was proven without the assumption in [37], together with the case $d = 2n - 5$ and all cases when $d \geq \frac{12n}{7}$.

We prove the conjecture for $d = 2n - 3$ and $d = 2n - 4$ following the proof in [37]. The case $d = 2n - 5$ is similar but much longer; the reader is referred to the outline in [37].

Lemma 19.1 *Suppose G is a regular graph with $2n$ vertices, other than K_{2n}. Then G has a one-factorization if and only if, for any vertex x of G, the graph $G' = G - x$ is of class 1.*

153

Proof. Suppose G is a regular graph of degree $d = \Delta(G)$.

First, assume G has a one-factorization. Then $\chi'(G) = d$, so obviously G' has a d-coloring. But since G is not complete it follows that there is some vertex y not adjacent to x in G, and y has degree d in G', so

$$\Delta(G') = d = \chi'(G').$$

So G' is class 1.

Now suppose G' is of class 1 for some x. Let π be a d-coloring of G'. Suppose $y_1, y_2, \cdots, y_d$ are the vertices of degree $d - 1$ in G' (the vertices adjacent to x in G). Since G' has $2n - 1$ vertices, each color class can contain at most $n - 1$ edges; and since G' has $(n - 1)d$ edges, it follows that each color class under π contains exactly $n - 1$ edges. So each color misses precisely one vertex. These vertices must be the y_i (since they are the vertices of degree less than d); if c_i is the color missing at y_i, it follows that the c_i are all different, so we can extend π to a d-coloring of G by the law $\pi(xy_i) = c_i$. So $\chi(G) = d$, and G has a one-factorization. $\square$

Theorem 19.2 *If G has $2n$ vertices and is regular of degree $2n - 3$, then G has a one-factorization.*

Proof. Suppose x is any vertex of G. Then $G - x$ has two vertices of degree $2n - 3$ and all other vertices of degree $2n - 4$. $G - x$ has only two major vertices, so from Corollary 8.10.2 it is class 1. So the Lemma ensures that G has a one-factorization. $\square$

Lemma 19.3 *Suppose e is an edge of the graph G and x is a vertex of G which is adjacent to at most one major vertex. Then:*

$$\Delta(G - e) = \Delta(G) \Rightarrow \chi'(G - e) = \chi'(G);$$
$$\Delta(G - x) = \Delta(G) \Rightarrow \chi'(G - x) = \chi'(G).$$

Proof. Either G is class 1 or class 2. If G is class 1, then in the first case

$$\Delta(G) = \chi'(G) \geq \chi'(G - e) \geq \Delta(G - e) = \Delta(G),$$

so $\chi'(G - e) = \Delta(G - e) = \chi'(G)$, and similarly for $G - x$. If G is class 2, we can treat both cases at once by taking x as an endpoint of e Let G' be

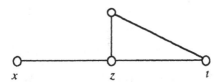

Figure 19.1   Adjacencies among major vertices.

a Δ-critical subgraph of G. By Corollary 8.10.1, x is not in G', so G' is a subgraph of $G - x$. So $G - x$ and $G - e$ are class 2: $\Delta(G - e) = \Delta(G)$ implies $\chi'(G - e) = \chi'(G) = \Delta$, and similarly for $G - x$.   □

Lemma 19.4 *Suppose G is a class 2 graph on v vertices, $v \geq 10$, which has exactly three major vertices. Then $\Delta(G) = v - 1$ and $\delta(G) = v - 2$.*

Proof. Suppose G is a class 2 graph with exactly three major vertices x, y and z. Say G' is a Δ-critical subgraph of G. Then G' has at least three major vertices, by Corollary 8.10.2. So it has precisely three major vertices, namely x, y and z. From Corollary 8.10.3, $\delta(G')$ is at least $\Delta - 1$, so $\delta(G') = \Delta - 1$.

Suppose there were a vertex u in G but not in G'. Then u cannot be adjacent to x, y or z (because the vertex adjacent to u would not be major in G'), nor to any other vertex (the vertex adjacent to u would have degree less than $\Delta-$ inG'). So no such u exists. So $V(G') = V(G)$, and therefore obviously $G' = G$.

Since $G = G'$, $\delta(G) = \Delta - 1$. Since G has 3 vertices of degree Δ and $v - 3$ vertices of degree $\Delta - 1$, the sum of its degrees is $v\Delta - v + 3$; since this must be even, v must be odd and Δ must be even. Say $v = 2p + 1$. By assumption $p \geq 5$.

We observe two facts. First, from Corollary 8.10.1, x, y and z are all adjacent. Second, from Corollary 8.10.4, $\Delta \geq \frac{2}{3}v$.

Assume the Lemma is false: then $\Delta \leq 2p - 2$. So there is some vertex t not adjacent to x. Then from $\delta(G) \geq \Delta - 1$ we see that in the graph $G - x - y$ formed by deleting x and y from G we have

$$
\begin{aligned}
\delta(G - x - y) &\geq \Delta - 3 \\
&\geq \tfrac{2}{3}(2p + 1) - 3
\end{aligned}
$$

$$\geq \quad \lceil \tfrac{1}{3}(4p + 2) \rceil - 3$$
$$\geq \quad \tfrac{1}{2}(2p - 1)$$
$$= \quad \tfrac{1}{2}|V(G - x - y)|,$$

so from Corollary 2.3.1, $G - x - y$ has a Hamilton cycle, say $t, t_1, t_2, \cdots, t_{2p-2}$. Define H to be the graph obtained from G by deleting the set of edges $E = \{t_1 t_2, t_3 t_4, \cdots, t_{2p-3} t_{2p-2}, xy\}$. Then H has four major vertices, x, y, z and t, of degree $\Delta - 1$, whose adjacencies are shown in Figure 19.1. Since x is adjacent to only one major vertex in H and clearly $\Delta(H - xz) = \Delta(H)$, it follows from Lemma 19.3 that $\chi'(H - xz) = \chi'(H)$. But $H - xz$ has only two major vertices, so, by Corollary 8.10.2, $H - xz$ is class 1. So $\chi'(H - xz) = \Delta(H - xz)$. Consequently $\chi'(H) = \Delta(H)$. Given a $\Delta(H)$-coloring of H, we can add a new color and apply it to every edge in E, forming a coloring of G in $\Delta(H) + 1 = \Delta(G)$ colors — a contradiction. $\qquad \square$

Lemma 19.5 *Every regular graph of degree $2n - 4$ on $2n$ vertices has a one-factorization provided $2n \geq 12$.*

Proof. Suppose G is a regular graph of degree $2n - 4$ on $2n$ vertices, where $2n \geq 12$, and x is a vertex of G. Then $G - x$ has exactly three major vertices, so by Lemma 19.4 it is class 1. Therefore Lemma 19.1 tells us that G is of class 1. $\qquad \square$

Lemma 19.6 *Every regular graph of degree $2n - 4$ on $2n$ vertices, $2n \leq 10$, has a one-factorization, except for $K_3 \cup K_3$.*

Proof. When $2n = 6$, the only graphs are C_6 and $K_3 \cup K_3$, and C_6 has a one-factorization. When $2n = 8$, there are six regular graphs of degree 4 — their complements are shown in Figure 19.2. It is left as an exercise to show that all six have one-factorizations.

There are 21 graphs for $2n = 10$; their complements are shown in Figure 19.3. (This list of graphs has been computed several times — see for example [28], whose labeling we use.) We prove that all of the graphs have one-factorizations.

We first consider the following one-factorization of K_{10}, which is $\mathcal{F}219$ in Appendix A:

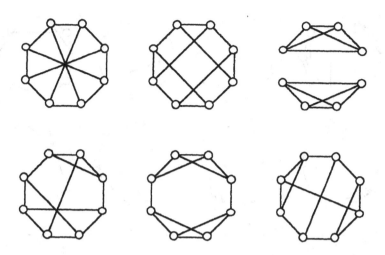

Figure 19.2 Cubic graphs on eight vertices.

$F_1 = 01,23,45,67,89$ $F_2 = 02,14,36,58,79$ $F_3 = 03,15,28,47,69$
$F_4 = 04,17,29,35,68$ $F_5 = 05,19,26,34,78$ $F_6 = 06,18,24,39,57$
$F_7 = 07,12,38,46,59$ $F_8 = 08,16,25,37,49$ $F_9 = 09,13,27,48,56$

Clearly the complement of any union of these factors has a one-factorization (just take the factors not used in the union). Now sixteen of the cubic graphs on ten vertices can be expressed as unions of three of the factors: see Table 19.1.

$G2 = F_2 \cup F_4 \cup F_9$ $G11 = F_1 \cup F_2 \cup F_5$
$G3 = F_1 \cup F_4 \cup F_5$ $G12 = F_1 \cup F_2 \cup F_9$
$G4 = F_3 \cup F_7 \cup F_8$ $G13 = F_1 \cup F_2 \cup F_3$
$G5 = F_1 \cup F_3 \cup F_4$ $G14 = F_3 \cup F_4 \cup F_9$
$G6 = F_1 \cup F_3 \cup F_7$ $G15 = F_2 \cup F_4 \cup F_8$
$G7 = F_1 \cup F_4 \cup F_8$ $G16 = F_1 \cup F_2 \cup F_5$
$G8 = F_1 \cup F_3 \cup F_5$ $G17 = F_1 \cup F_5 \cup F_8$
$G9 = F_1 \cup F_2 \cup F_7$ $G18 = F_1 \cup F_3 \cup F_8$

Table 19.1 Factoring cubic graphs on ten points.

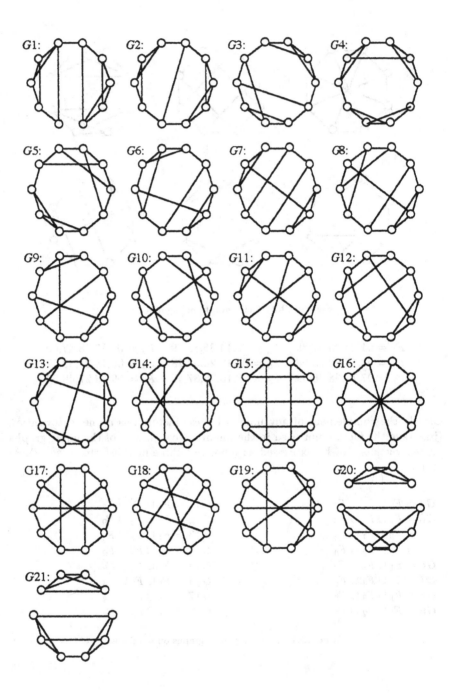

Figure 19.3   Cubic graphs on ten vertices.

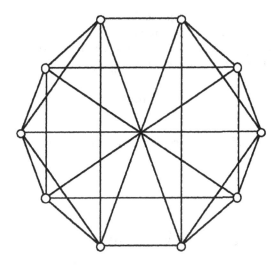

Figure 19.4 Graph containing $G1$, $G20$ and $G21$.

Graph $G19$ is the Petersen graph; we leave it to the reader (Exercise 19.2) to prove that its complement has a one-factorization. Graph $G10$ is bipartite, and it is easy to prove that its complement has a one-factorization (see Exercise 19.3).

We are left with $G1$, $G20$ and $G21$. We observe that each of these is a subgraph of the graph H shown in Figure 19.4, and that the complement in H of each of the graphs is a union of two one-factors. On the other hand, $\overline{H}$ is the union of the one-factors

$$02, 14, 37, 58, 69 \qquad 03, 17, 28, 46, 59$$
$$04, 18, 26, 39, 57 \qquad 07, 19, 25, 36, 48.$$

Therefore each of the complements has a one-factorization.

From Lemmas 19.5 and 19.6 we have

Theorem 19.7 *Every regular graph of degree $2n - 4$ on $2n$ vertices has a one-factorization, other than $K_3 \cup K_3$. So the one-factorization conjecture is true when $d = 2n - 4$.* □

We now prove the bound $(\sqrt{7} - 1)n$, following [39]. If x and y are vertices of a graph G, we write $p_{xy}(G)$ for the number of paths of length 2 from x to y in G, and $\bar{p}_{xy}(G) = p_{xy}(\overline{G})$, so that $\bar{p}_{xy}(G)$ the number of vertices of G adjacent neither to x nor to y. Define $\bar{p} = \bar{p}(G) = \max_{x,y} \bar{p}_{xy}(G)$. Clearly

$$\bar{p} \leq \Delta(\overline{G}) = |V(G)| - d(G) - 1.$$

Lemma 19.8 [39]   *If G is a regular graph on $2n$ vertices whose degree d satisfies*

$$d \geq \tfrac{5}{3}n - \tfrac{1}{3}\bar{p} - \tfrac{1}{6}$$

then G has a one-factorization.

Proof. Let G be a regular graph on $2n$ vertices with degree d, where

$$d \geq \tfrac{5}{3}n - \tfrac{1}{3}\bar{p} - \tfrac{1}{6}.$$

Let w and v^* be vertices of G such that the number of paths of length two between w and v^* in $\overline{G}$ is $\bar{p}$. Let W be the set of vertices of maximum degree in G. Then $G - w$ has $2n - 1$ vertices, $|W| = 2n - d - 1$ of which have degree d, and the remaining d have degree $d - 1$. The vertex v^* is non-adjacent to $\bar{p}$ of the vertices of W. Thus $d(v^*)$ equals either $|W| - \bar{p}$ or $|W| - \bar{p} - 1$.

Let X be a set of $|W| - \bar{p} - 1$ vertices of $V(G - w)$ which are not adjacent to v^*; as there are in $G - w$ at least $|W| - 1$ vertices not adjacent to v^*, such a set X does exist. Write $s = |X| = |W| - \bar{p} - 1$, and write $q = |(X \cup W)\backslash\{v^*\}|$. Then

$$q \leq (|W| - \bar{p} - 1) + |W| = 2|W| - \bar{p} - 1 = 4n - 2d - \bar{p} - 3.$$

Now consider the subgraph H of $G - w$ induced by $(X \cup W)\backslash\{v^*\}$. Let M_0 be a set of edges of H forming a maximal matching in H; write $m = |M_0|$. Consider a decomposition

$$(X \cap M)\backslash\{v^*\} = L \cup R,$$

where $L \cap R = \emptyset$, $|L| = m$, in which every edge of M_0 joins a vertex of L to a vertex of R. Write L^* for $L \cup \{v^*\}$ and let H^* be the subgraph of $G - \{w\}$ induced by $L^* \cup R$. Let the elements of L^* be denoted by $\ell_1, \cdots, \ell_{m+1}$, where $\ell_{m+1} = v^*$, and let the elements of R be denoted by $r_t, r_{t+1}, \cdots, r_m$, where $t = 2m - q + 1$ (t may be negative). Without loss of generality, we may assume that

$$M_0 = \{\ell_1 r_1, \ell_2 r_2, \cdots, \ell_m r_m\}.$$

Let E^+ consist of all edges of H^*, except for those of the form $\ell_i r_j$, $i \leq j$, and those with both endpoints in R.

Observe that E^+ is contained in the union of q edge-disjoint matchings of the complete graph on $V(H^*)$, $M_1^+, \cdots, M_q^+$, where M_i^+ is defined as follows:

$$M_i^+ = \{\ell_1 r_{1-i}, \ell_2 r_{2-i}, \cdots, \ell_m r_{m-i}, \ell_{m+1} r_{m+1-i}\}, \quad \text{if } 1 \leq i \leq 1 - t,$$

$$M_i^+ = \{\ell_1 \ell_{i+t-1}, \ell_2 \ell_{i+t-2}, \cdots, \ell_{\lfloor(i+t-1)/2\rfloor}\ell_{\lceil(i+t-1)/2\rceil+1}\}$$
$$\cup \{\ell_{i+t} r_t, \ell_{i+t+1} r_{t+1}, \cdots, \ell_m r_{m-i}, \ell_{m+1} r_{m+1-i}\},$$
$$\text{if } 2 - t \leq i \leq m + 1 - t,$$

$$M_i^+ = \{\ell_{i-m+t-1}\ell_{m+1}, \ell_{i-m+t}\ell_m, \cdots, \ell_{\lfloor(i+t-1)/2\rfloor}\ell_{\lceil(i+t-1)/2\rceil+1}\},$$
$$\text{if } m + 2 - t \leq i \leq q.$$

Notice that $\bigcup_{i=1}^{q} M_i^+$ contains all the edges of the complete graph on the vertices of $V(H^*)$ except for the edges of M_0, the edges with both endpoints in R, and the edges which join $\ell_i \in L$ to $r_j \in R$ with $i < j$. Finally observe that

$$|M_i^+| \leq \tfrac{1}{2}(q + 2 - i) \quad (1 \leq i \leq q).$$

Let W_s be a set of s elements of W which are adjacent to v^*. (Recall that there are either s or $s + 1$ such elements.) Let the vertices of X be $x_1, x_2, \cdots, x_s$ and the vertices of W_s be $w_1, w_2, \cdots, w_s$. If an edge xw is in M_0 with $x \in X$ and $w \in W_s$, we may assume that $x \in R$ and $w \in L$. We may further assume that $\ell_1, \cdots, \ell_m, r_1, \cdots, r_m, x_1, \cdots, x_s, w_1, \cdots, w_s$ are labelled so that, for $1 \leq j \leq s - 1$, x_j comes before w_j in the list $(r_m, \cdots, r_t, \ell_1, \cdots, \ell_m)$. (In fact, except in the case when each edge of M_0 joins either two vertices of X or two vertices of W_s, we could assume that x_s comes before w_s also.)

We now construct matchings M_i^* $(1 \leq i \leq q + 1)$ by modifying the M_i^+. If x_s comes before w_s, define $M_i^* = M_i^+$ $(1 \leq i \leq q)$ and $M_{q+1}^* = \emptyset$. If x_s comes after w_s, then we may assume that $v^* w_s \in M_{i_0}^+$ for some i_0. If x_0 is not incident with any edge in $M_{i_0}^+$, then define $M_i^* = M_i^+$ $(1 \leq i \leq q)$ and $M_{q+1}^* = \emptyset$. If there is an edge in $M_{i_0}^+$ incident with x_0, say e_{i_0}, then define $M_i^* = M_i^+$ if $i \in \{1, \cdots, q\} \setminus \{i_0\}$, $M_{i_0}^* = M_{i_0} \setminus \{e_{i_0}\}$ and $M_{q+1}^* = \{e_{i_0}\}$.

Notice that

$$|M_k^*| \leq \tfrac{1}{2}(q + 3 - k) \quad (1 \leq k \leq q + 1).$$

Next we shall choose $q + 1$ edge-disjoint near-one-factors $F_1, \cdots, F_{q+1}$ of $G - w$ such that

$$E^+ \cap (M_1^* \cup \cdots \cup M_k^*) \subseteq F_1 \cup \cdots \cup F_k \quad (1 \le k \le q+1),$$
$$M_0 \cap (F_1 \cup \cdots \cup F_{q+1}) = \emptyset$$

and furthermore,

> if $v^*w_i \in M_k^*$ for some $i \in \{1, 2, \cdots, s\}$,
>
> then $v^*w_i \in F_k$ and F_k omits vertex x_i,

and

> if $v^*w_i \notin M_k$ for all $i \in \{1, 2, \cdots, s\}$, then F_k omits v^*.

To construct F_k $(1 \le k \le q+1)$, suppose that $F_1, \cdots, F_{k-1}$ have been chosen already. Write

$$M_k = (E^+ \cap M_k^*) \backslash (F_1 \cup \cdots \cup F_{k-1} \cup M_0).$$

Then

$$|M_k^*| \le \tfrac{1}{2}(q+3-k) \quad (1 \le k \le q+1).$$

Consider

$$G_{k-1} = (G - w) \backslash (F_1 \cup \cdots \cup F_{k-1} \cup M_0).$$

We take F_k to be a near-one-factor of G_{k-1} containing M_k and missing x_i if $v^*w_i \in M_k$ for some $i \in \{i, \cdots, s\}$, or missing v^* if $v^*w_i \notin M_k$ for all $i \in \{i, \cdots, s\}$.

Let $V(M_k)$ denote the set of vertices of G which are incident with the edges of M_k, and define G_{k-1}^* by $G_{k-1}^* = G_{k-1} \backslash V(M_k)$. To see that we can select an F_k as described, we apply Corollary 2.3.1 (Dirac's Theorem) to show that G_{k-1}^* has a Hamilton cycle. We have

$$\delta(G_{k-1}^*) \ge (d-1) - \{(k-1)+1\} - |V(M_k)| = d - k - 1 - |V(M_k)|.$$

Also

$$\begin{aligned}
\tfrac{1}{2}|V(G_{k-1}^*)| &= \tfrac{1}{2}\{|V(G_{k-1})| - |V(M_k)|\} \\
&= \tfrac{1}{2}\{2n - 1 - |V(M_k)|\} \\
&= n - \tfrac{1}{2} - \tfrac{1}{2}|V(M_k)|.
\end{aligned}$$

Therefore

$$\delta(G_{k-1}^*) - \tfrac{1}{2}|V(G_{k-1}^*)| \ge d - k - 1 - |V(M_k)| - n + \tfrac{1}{2} + \tfrac{1}{2}|V(M_k)|$$

$$
\begin{aligned}
&= & d - n - k - \tfrac{1}{2} - \tfrac{1}{2}|V(M_k)| \\
&\geq & d - n - k - \tfrac{1}{2} - \tfrac{1}{2}(q + 3 - k) \\
&= & d - n - \tfrac{1}{2}q - \tfrac{1}{2}k - 2 \\
&\geq & d - n - q - \tfrac{5}{2} \\
&\geq & d - n - (4n - 2d - \bar{p} - 3) - \tfrac{5}{2} \\
&= & 3d - 5n + \bar{p} + \tfrac{1}{2} \\
&\geq & 0,
\end{aligned}
$$

since $d \geq \tfrac{5}{6}(2n) - \tfrac{1}{3}\bar{p} - \tfrac{1}{6}$. Therefore by Corollary 2.3.1, G_{k-1}^* does have a Hamilton cycle. It follows that G_{k-1} contains a near-one-factor F_k which contains M_k and misses x_i if $v^*w_i \in M_k^*$ for some $i \in \{1\cdots,s\}$, or misses v^* if $v^*w_i \notin M_k^*$ for all $i \in \{1,\cdots,s\}$. It is easy now to check that F_k has all the properties required of it.

Let $J = \{i : F_i \text{ omits } v^*\}$. The graph $((G - w)\backslash(F_1 \cup \cdots \cup F_{q+1})) - \{v^*\}$ has core of the form discussed in Lemma 8.11. So it is class 1. So

$$
\begin{aligned}
&(((G - w)\backslash(F_1 \cup \cdots \cup F_{q+1})) - \{v^*\}) \cup \{F_i : i \in J\} \\
&= ((G - w)\backslash\{F_i : i \in \{1, \cdots, q + 1\}\backslash J\}) - \{v^*\}
\end{aligned}
$$

is also class 1. By Corollary 8.8.10.5, it now follows that

$$
(G - w)\backslash\{F_i : i \in \{1, \cdots, q + 1\}\backslash J\}
$$

is class 1. It therefore follows that $G - w$ is class 1. In any edge-coloring of $G - w$ with $d(G)$ colors, it is easy to see by counting that each color is missing from exactly one vertex. Therefore an edge-coloring of $G - w$ can be extended to an edge-coloring of G. Thus G is class 1. $\qquad\square$

We next need a bound on $\bar{p}$.

Lemma 19.9 [39] *If G has $2n$ vertices and is regular of degree d, then*

$$
\bar{p}(G) \geq \frac{(2n - d - 1)(2n - d - 2)}{2n - 1}.
$$

Proof. Each vertex in $\overline{G}$ is the center of $\binom{2n-d-1}{2}$ paths of length 2 in $\overline{G}$. So the total number of such paths is $2n\binom{2n-d-1}{2}$. Averaging between all $\binom{2n}{2}$

possible pairs of endpoints, we see that the average value of $\bar{p}_{xy}$, for all pairs $\{x, y\}$, is

$$\frac{2n \binom{2n-d-1}{2}}{\binom{2n}{2}} = \frac{(2n - d - 1)(2n - d - 2)}{2n - 1},$$

and surely $\bar{p}$ is at least as large as the average.   □

Theorem 19.10 [39] *If G is a regular graph of degree d on $2n$ vertices, then G has a one-factorization if*

$$d \geq (\sqrt{7} - 1)n.$$

Proof. From Lemma 19.8, we know that G has a one-fac torization provided

$$
\begin{aligned}
d &\geq \tfrac{5}{3}n - \tfrac{1}{3}\bar{p} - \tfrac{1}{6} \\
&\geq \frac{5}{3}n - \frac{(2n - d - 1)(2n - d - 2)}{3} - \frac{1}{6},
\end{aligned}
$$

from Lemma 19.9. After simplifying, this inequality becomes

$$d^2 + 2nd - (6n^2 - \tfrac{3}{2}) \geq 0,$$

or equivalently

$$(d + n)^2 \geq 7n^2 - \tfrac{3}{2}.$$

So G has a one-factorization if

$$d \geq (\sqrt{7} - 1)n.$$ □

Exercises 19

19.1 Prove that every connected regular graph of degree 3 on 8 vertices is a union of three factors in the factorization $\mathcal{F}_5$ of K_8, as presented in Table 11.11.1. Hence, prove that every regular graph of degree 4 on 8 vertices has a one-factorization, completing the proof of Lemma 19.6.

19.2 Prove that the complement of the Petersen graph is a union of three edge-disjoint Hamilton cycles, and consequently has a one-factorization.

19.3 The graph G consists of two copies of K_n, together with a one-factor each edge of which joins a member of one K_n to a member of the other. Prove that G has a one-factorization. Hence prove that any regular bipartite graph on $2n$ vertices, whose degree is less than n, has a one-factorization.

PREMATURE SETS OF FACTORS

Suppose S is a set of compatible one-factors of the regular graph G which cannot be embedded in a one-factorization of G. Then S is called a *premature set of factors* of G. Any proper maximal set is premature, but not conversely; if fact we are most interested in what we might call *properly premature* sets — sets which can be extended part way, but not to a one-factorization. We discuss premature and properly premature sets in complete graphs.

Theorem 20.1 [132] *There is a premature set of k one-factors in K_{2n} whenever k is even and $n < k < 2n - 4$, or when $k = 2n - 4, n \geq 5$ and n is odd.*

Proof. Let $U = \{1, 2, \cdots, 2r + 1\}, \overline{U} = \{\overline{1}, \overline{2}, \cdots, \overline{2r+1}\}$. Consider the complete bipartite graph $K_{2r+1,2r+1}$ with bipartition $(U, \overline{U})$. Let $L = \{L_0, L_1, \cdots, L_{2r}\}$ be a one-factorization of this graph, where

$$L_i = \big\{ \{1, \overline{1+i}\}, \{2, \overline{2+i}\}, \cdots, \{2r+1, \overline{i}\} \big\}$$

(the labels being reduced modulo $2r + 1$ whenever necessary).

If n is odd, say $n = 2r + 1$, take F_i to be L_i for $i = 1, 2, \cdots, 2r$. Since $k > n$ and k is even, we can write $k = 2r + s$ for some positive even integer s. Let now $E = \{E_1, E_2, \cdots, E_{2r+1}\}$ be a one-factorization of the K_{2r+2} on $\{\infty, 1, 2, \cdots, 2r+1\}$ with the property that $\{\infty, i\} \in E_i$; write $\overline{E}_i$ for E_i with x replaced by $\overline{x}$ for each vertex x. Without loss of generality we may assume that E_s contains edges $(s+1, s+2), (s+3, s+4), \cdots, (2r-1, 2r)$. Define

$$F_{2r+i} = E_i \cup \overline{E}_i \cup \big\{ (i, \overline{i}) \big\} \setminus \big\{ (\infty, i), (\overline{\infty}, \overline{i}) \big\}, \quad 1 \leq i \leq s - 1$$

$$F_{2r+s} = E_s \cup \overline{E}_s \cup \left\{ (s, \overline{s}), (s+1, \overline{s+1}), \cdots, (2r, \overline{2r}) \right\}$$
$$\setminus \left\{ (\infty, s), (\overline{\infty}, \overline{s}), (s+1, s+2), (\overline{s+1}, \overline{s+2}), \right.$$
$$\left. \cdots, (2r-1, 2r), (\overline{2r-1}, \overline{2r}) \right\}.$$

Then $F_1, F_2, \cdots, F_{2r+s}$ is a compatible set of k one-factors of the K_{2n} on $\{1, 2, \cdots, 2r+1, \overline{1}, \overline{2}, \cdots, \overline{2r+1}\}$ but its leave has no one-factorization, because it contains $(2r+1, \overline{2r+1})$ as a bridge.

Suppose n is even: $n = 2r$. Take F_i to be

$$L_i \cup \left\{ (2r-i, 2r-i+1) \right\} \setminus \left\{ (2r-i, \overline{2r}), (2r-i+1, \overline{2r+1}) \right\}$$

which is a one-factor in a graph with vertices $\{1, 2, \cdots, 2r+1, \overline{1}, \overline{2}, \cdots, \overline{2r-1}\}$. Again, write $k = 2r + s$. Let $E = \{E_1, E_2, \cdots, E_{2r+1}\}$ be a one-factorization of the K_{2r+2} on $\{\infty, 1, 2, \cdots, 2r+1\}$ with the properties that E_i contains (∞, i) for all i, and that E_s contains $(s+1, s+t+1), (s+2, s+t+2), \cdots, (s+t, s+2t)$ and $(2r, 2r+1)$ where $t = r - 1 = \frac{s}{2}$. Let further $D = \{D_1, D_2, \cdots, D_{2r-1}\}$ be a one-factorization of the K_{2r} on $\{\overline{\infty}, \overline{1}, \overline{2}, \cdots, \overline{2r-1}\}$ with the properties that $(\overline{\infty}, \overline{i}) \in D_i$ for all i and that D_s contains $(\overline{s+1}, \overline{s+t+1}), (\overline{s+2}, \overline{s+t+2}), \cdots, (\overline{s+t}, \overline{s+2t})$. Define now

$$F_{2r+i} = E_i \cup D_i \cup \left\{ (i, \overline{i}) \right\} \setminus \left\{ (\infty, i), (\overline{\infty}, \overline{i}) \right\}, \quad 1 \le i \le s-1$$
$$F_{2r+s} = E_s \cup D_s \cup \left\{ (s, \overline{s}), (s+1, \overline{s+1}), \cdots, (s+2t, \overline{s+2t}) \right\}$$
$$\setminus \left\{ (\infty, s), (\overline{\infty}, \overline{s}), (s+1, s+t+1), (\overline{s+1}, \overline{s+t+1}), \right.$$
$$\left. \cdots, (s+2t-1, s+2t), (\overline{s+2t-1}, \overline{s+2t}) \right\}.$$

The one-factors $F_1, F_2, \cdots, F_{2r+s}$ are compatible provided $t > 1$, i.e. $s < 2r - 4$, and they form a premature set because $(2r-1, \overline{2r-1})$ is a bridge in its leave. $\square$

The above result, together with the existence of maximal sets as outlined in Chapter 18, proves the existence of premature sets in all cases where $n < k \le 2n - 3$, or when $n = k$ and n is even, except for the single case of $k = 8, 2n = 12$. However, the factors

$$
\begin{array}{rcllllll}
F_1 & = & 01 & 28 & 39 & 4a & 5b & 67 \\
F_2 & = & 02 & 16 & 35 & 4b & 78 & 9a \\
F_3 & = & 05 & 18 & 29 & 3a & 4b & 7b \\
F_4 & = & 07 & 12 & 38 & 49 & 5a & 6b \\
F_5 & = & 08 & 19 & 2a & 3b & 47 & 56 \\
F_6 & = & 09 & 1a & 2b & 37 & 45 & 68 \\
F_7 & = & 0a & 1b & 27 & 34 & 58 & 69 \\
F_8 & = & 0b & 17 & 23 & 48 & 59 & 6a \\
\end{array}
$$

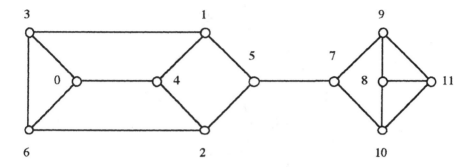

Figure 20.1 This graph has no one-factorization.

are compatible, and their leave (shown in Figure 20.1) contains the bridge 57, so it has no one-factorization. From this we deduce:

Corollary 20.1.1 K_{2n} *contains a premature set of one-factors whenever* $n \le k \le 2n - 3$, n *odd, or* $n < k \le 2n - 3$, n *even.* $\qquad\square$

To discuss properly premature sets we use the following recursive construction. (Again twin factorizations are involved.)

Theorem 20.2 [163] *Suppose there is a properly premature set of* k *one-factors in* K_{2n}. *Then there are properly premature sets of* $2n + k$ *factors in* K_{4n} *and* $2n + k - 2$ *factors in* K_{4n-2}.

Proof. Suppose the K_{2n} with vertices $1, 2, \cdots, 2n$ contains pairwise disjoint one-factors $F_1, F_2, \cdots, F_k$, and that this set can be extended but not completed. Select any k pairwise compatible one-factors in the K_{2n} with vertices $2n+1, 2n+2, \cdots, 4n$; call them $G_1, G_2, \cdots, G_k$. Then a suitable set consists of the factors $F_1 \cup G_1, F_2 \cup G_2, \cdots, F_k \cup G_k$, together with the $2n$ factors in a one-factorization of the $K_{2n,2n}$ with vertex sets $\{1, 2, \cdots, 2n\}$ and $\{2n+1, 2n+2, \cdots, 4n\}$.

A similar approach for K_{4n-2} fails, since we cannot factor $K_{2n,2n-2}$. Instead, we first choose $k - 2$ compatible one-factors of K_{2n-2}, say $G_3, G_4, \cdots, G_k$, and take

$$F_3 \cup G_3, F_4 \cup G_4, \cdots, F_k \cup G_k$$

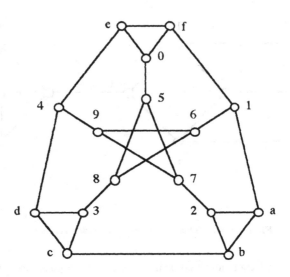

Figure 20.2   Another graph with no one-factorization.

as $k-2$ of our factors. The other $2n$ factors constitute a one-factorization of $F_1 \cup F_2 \cup K_{2n,2n-2}$. It remains to show that such a factorization is possible.

Suppose we could find a one-factorization $H_1, H_2, \cdots, H_{2n}$ of $K_{2n,2n}$ with the following property: for $i = 1, 2, \cdots, 2n$, factor H_i contains the edges $(x_i, 4n-1)$ and $(y_i, 4n)$, where the $2n$ edges (x_i, y_i) together constitute $F_1 \cup F_2$. Then take J_i to be H_i with $(x_i, 4n-1)$ and $(y_i, 4n)$ deleted and replaced by (x_i, y_i). The factors $J_1, J_2, \cdots, J_{2n}$ constitute a one-factorization of $F_1 \cup F_2 \cup K_{2n,2n-2}$, as required.

But such a factorization exists. Since $F_1 \cup F_2$ is a regular graph of degree 2, it is a union of disjoint even cycles, say

$$(x_1 \, x_2 \, \cdots \, x_k) \cup (x_{k+1} \, x_{k+2} \, \cdots \, x_\ell) \cup \cdots \cup (\cdots \, x_{2n}).$$

Write $y_1 = x_2, \cdots, y_k = x_1$; $y_{k+1} = x_{k+2}, \cdots, y_\ell = x_{k+1}$; and so on. We need to construct a one-factorization $H_1, H_2, \cdots, H_{2n}$ of $K_{2n,2n}$ in which $(x_i, 4n-1)$ and $(y_i, 4n)$ belong to H_i for each i. This is easily done (see Exercise 18.1). So the required one-factorization of K_{4n-2} exists.   □

Corollary 20.2.1 *For every even order $2n$ greater than 8 there is a properly premature set of $2n-4$ one-factors in K_{2n}.*

Proof. From the Theorem, it is sufficient to prove the existence of such a set in K_{10}, K_{12}, K_{14} and K_{16}. Corollary 20.1.1 proves the existence of premature sets in all these cases. It follows from Theorem 18.3 that none of them can be maximal except in the case of K_{16}. To settle that case, consider the graph G in Figure 20.2. It contains one-factors; but, for example, the only factors containing ab are

$$ab \quad cd \quad ef \quad 05 \quad 16 \quad 27 \quad 38 \quad 49$$
$$ab \quad c3 \quad d4 \quad e0 \quad f1 \quad 27 \quad 58 \quad 69$$

and it is easy to check that neither can be extended to a one-factorization of G. On the other hand $K_{16}\backslash G$ has a one-factorization; in fact $K_{16}\backslash G$ can be written as a union of six Hamilton cycles:

$$(0\ 1\ e\ 2\ f\ 8\ c\ 9\ a\ 5\ d\ 7\ 3\ 6\ b\ 4)$$
$$\cup \quad (0\ 9\ 3\ a\ f\ 5\ b\ 3\ 4\ 1\ d\ 8\ 2\ 6\ c\ 7)$$
$$\cup \quad (0\ c\ f\ 9\ d\ a\ 8\ b\ e\ 6\ 7\ 4\ 5\ 1\ 2\ 3)$$
$$\cup \quad (0\ a\ 3\ e\ 7\ b\ f\ 4\ c\ 1\ 8\ 9\ 5\ 6\ d\ 2)$$
$$\cup \quad (0\ 6\ f\ 3\ 9\ 2\ 4\ a\ c\ 5\ e\ d\ b\ 1\ 7\ 8)$$
$$\cup \quad (0\ b\ 9\ 1\ 3\ 5\ 2\ c\ e\ 8\ 4\ 6\ a\ 7\ f\ d).$$

So the set of twelve one-factors making up these cycles is properly premature.

□

Exercises 20

20.1   (i) Suppose $L = (\ell_{ij})$ is a Latin square of side n. Prove that if

$$F_i = \left\{(x, y + n) : \ell_{xy} = i\right\}$$

then $\{F_1, F_2, \cdots, F_n\}$ form a one-factorization of the $K_{n,n}$ with vertex sets $\{1, 2, \cdots, n\}$ and $\{n + 1, n + 2, \cdots, 2n\}$.

(ii) A *transversal* of a square array is a set of positions which contain exactly one representative of each row, and exactly one representative of each column. (One obvious example is the diagonal.) Suppose k pairwise disjoint transversals are chosen in an $n \times n$ array; every position in the first transversal is filled with a 1, every cell in the second with a 2, and so on. Use part (i) to prove that, after this is done, the symbols $k + 1, k + 2, \cdots, n$ can be placed in each row so as to form a Latin square.

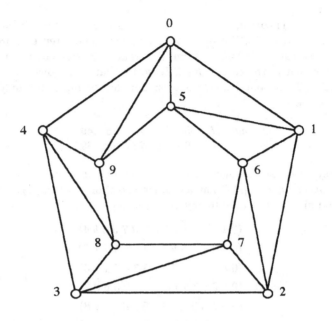

Figure 20.3   Graph for Exercise 20.2.

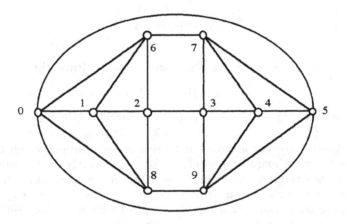

Figure 20.4   Graph for Exercise 20.4.

(iii) Hence prove the existence of the one-factorization required at the end of the proof of Theorem 20.2.

20.2 In the graph of Figure 20.3, verify that the one-factors $\{05, 16, 27, 38, 49\}$ and $\{09, 15, 26, 37, 48\}$ are compatible but cannot be completed.

20.3 Does the graph of Figure 20.3 have a one-factorization?

20.4 Find a pair of one-factors in the graph shown in Figure 20.4 which have the same property as those given in Exercise 20.2. Does this graph have a one-factorization?

21

CARTESIAN PRODUCTS

Various techniques have been studied which produce a new graph from two given graphs. There is some interest in the following problem: given a form of graph product, what conditions on graphs G and H imply that the product of G and H has a one-factorization? The question has been studied for cartesian products, wreath products and tensor products, among others.

The *cartesian product* $G \times H$ of graphs G and H is defined as follows:

(i) label the vertices of H in some way;

(ii) in a copy of G, replace each vertex of G by a copy of H;

(iii) add an edge joining vertices in two adjacent copies of H if and only if they have the same label.

In other words, if G has vertex-set $V(G) = \{a_1, a_2, \cdots, a_g\}$ and H has vertex-set $V(H) = \{b_1, b_2, \cdots, b_h\}$, then $G \times H$ has vertex-set $V(G) \times V(H)$, and (a_i, b_j) is adjacent to (a_k, b_ℓ) if and only if *either* $i = k$ and b_j is adjacent in H to b_ℓ *or* $j = \ell$ and a_i is adjacent in G to a_k. It is clear that $G \times H$ and $H \times G$ are isomorphic. Similarly $(G \times H) \times J$ and $G \times (H \times J)$ are isomorphic, so one can define cartesian products of three or more graphs in a natural way.

One way of representing a cartesian product is to write the vertices in a rectangular lattice, where (a_i, b_j) occurs in row i and column j. Then each column is a copy of G and each row is a copy of H. For this reason it is convenient to refer to edges of the form $(a_i, b_k)(a_j, b_k)$ as *vertical* edges, and edges coming from copies of H as *horizontal* edges. An illustration is given in Figure 21.1.

If d_G and d_H are the degree functions in G and H respectively, then (a_i, b_j) has degree $d_G(a_i) + d_H(b_j)$. So $G \times H$ is regular if and only if both G and

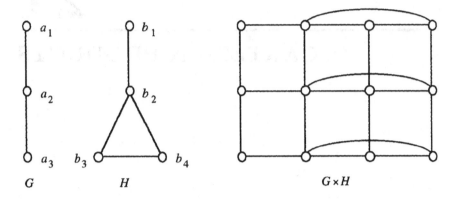

Figure 21.1   The cartesian product of two graphs.

H are regular. Thus, when thinking about one-factorizations, we need only consider the case where G and H are regular graphs; and of course at least one of them must have an even number of vertices. We shall see that these necessary conditions are not sufficient, but on the other hand G and H need not both have one-factorizations.

We start with some sufficient conditions.

Theorem 21.1 [111]   *If G is a class 1 graph other than a null graph then $G \times H$ is a class 1 graph.*

Proof. We know that $\chi'(G) = \Delta(G)$; by Theorem 8.4 either $\chi'(H) = \Delta(H)$ or $\chi'(H) = \Delta(H) + 1$. Select an edge-coloring of G in the $\Delta(G)$ colors $1, 2, \cdots, \Delta(G)$, and apply it to every copy of G in an identical way (that is, if edge $a_i a_j$ has color c, then every edge $(a_i, b_k)(a_j, b_k)$ in $G \times H$ receives color c). Select a coloring of H in colors $\Delta(G) + 1, \Delta(G) + 2, \cdots, \Delta(G) + \chi'(H)$, and use it to color the horizontal edges. Then we have a coloring of $G \times H$ in $\Delta(G) + \chi'(H)$ colors.

If $\chi'(H) = \Delta(H)$, we are done. If not, $\chi'(H) = \Delta(H) + 1$. We show how to eliminate one color, color 1 say. Choose any edge $a_i a_j$ of G such that $a_i a_j$ receives color 1. Given k, there will be one color c from $\{\Delta(G) + 1, \Delta(G) + 2, \cdots, \Delta(G) + \chi'(H)\}$ such that no edge of color c touches b_k in the coloring of

H. Recolor $(a_i, b_k)(a_j, b_k)$ in color c. The result is a $(\Delta(G) + \Delta(H))$-coloring. $\square$

Corollary 21.1.1 *If G is a non-null graph with a one-factorization and H is a regular graph, then $G \times H$ has a one-factorization.* $\square$

The Corollary was proven directly in [88] and [106]. The requirement that G be non-null is necessary; if G is null then $G \times H$ clearly has the same edge-chromatic number and class as H, and will have a one-factorization if and only if H does.

Before proving the next Theorem, we make the following observation.

Lemma 21.2 *Suppose $G = G_1 \cup G_2$ and $H = H_1 \cup H_2$ are edge-disjoint decompositions of G and of H into two spanning subgraphs. Then*

$$G \times H = (G_1 \times H_1) \cup (G_2 \times H_2)$$

and $G_1 \times H_1$ is edge-disjoint from $G_2 \times H_2$. $\square$

Theorem 21.3 [106] *If G and H are regular graphs and each contains a one-factor, then $G \times H$ has a one-factorization.*

Proof. (after [75]) Suppose G has a one-factor E and H has a one-factor F. Write

$$\begin{aligned} G_1 &= G - E, \\ H_1 &= H - F. \end{aligned}$$

Then

$$\begin{aligned} G \times H &= (E \cup G_1) \times (F \cup H_1) \\ &= (E \times H_1) \cup (F \times G_1), \end{aligned}$$

all the unions being edge-disjoint, by Lemma 21.2. Now Corollary 21.1.1 tells us that both $E \times H_1$ and $F \times G_1$ have one-factorizations, so $G \times H$ has a one-factorization. $\square$

The above results give sufficient conditions for a one-factorization, but they are not necessary. We shall soon prove that, if G is any cubic graph and H is a cycle of length at least 4, then $G \times H$ has a one-factorization. For example, let N be the graph of Figure 3.1. Then N has no one-factor but $N \times C_5$ has a one-factorization.

In order to prove this Theorem we need a new concept. We define a *Kotzíg coloring* of a cubic graph to be an edge-coloring in five colors $\{1, 2, 3, 4, 5\}$ such that the three edges incident with any vertex are colored with one of the combinations $\{1, 2, 3\}$, $\{2, 3, 4\}$, $\{3, 4, 5\}$, $\{4, 5, 1\}$ and $\{5, 1, 2\}$.

Lemma 21.4 [106] *If G is a cubic graph with a one-factor then G has a Kotzíg coloring.*

Proof. We suppose G has a one-factor F, and write Q for the complement of F in G. Q is a disjoint union of cycles. We write $\mathcal{C}_o(F)$ and $\mathcal{C}_e(F)$ for the sets of odd and even cycles making up Q respectively; say

$$\mathcal{C}_o(F) = \{C_1, C_2, \cdots, C_n\}.$$

Assume that F has been chosen so that n is minimized.

If $\mathcal{C}_o$ is empty there is nothing to prove, since G has a one-factorization and any 3-coloring is trivially a Kotzíg coloring; so we can assume $n \geq 1$. Select an edge h_i from each C_i and write $H = \{h_1, h_2, \cdots, h_n\}$; denote the path $C_i - h_i$ by P_i.

The subgraph $F \cup H$ has some vertices of degree 1 and some of degree 2, so it is a union of disjoint paths and cycles. Suppose S is one of the cycles (if there are any). Then S consists of alternating edges from H and F, so its length is even. Without loss of generality we can assume that the edges of H in S are $h_1, h_2, \cdots, h_s$. Now suppose we define a one-factor F' by

$$F' = F - (F \cap S) + (H \cap S).$$

For this factor we see that the cycles $C_1, C_2, \cdots, C_s$ in Q have been replaced by the single cycle

$$(S \cap L) \cup P_1 \cup P_2 \cup \cdots \cup P_s.$$

Since s must be at least 2 for S to be a cycle, we have at least one fewer odd cycle in this case than in the case of F; but this contradicts the minimality for which we chose F. So we can assume that $H \cup F$ is a union of p disjoint paths $R_1, R_2, \cdots, R_p$. Clearly the end edges in the paths will be members of F (any

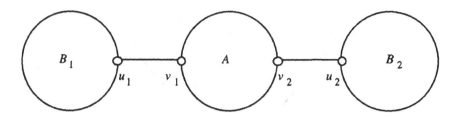

Figure 21.2 Putative graph with no Kotzíg coloring.

vertex belonging to a member of H will have degree 2 in $H \cup F$), so the paths are of odd length.

We now set up a coloring π. All the edges in H are colored with color 3, and those edges in each R_i which belong to F are colored alternately in colors 2 and 4. (As we go along R_i, the edges are colored in sequence 2, 3, 4, 3, 2, 3, 4, 3, 2, $\cdots$.) The edges of each member of C_e are colored alternately 1, 5, 1, 5, $\cdots$. To color the edges of the P_i, we observe that one endpoint of P_i touches edges of colors 2 and 3, and the other touches edges of colors 3 and 4 (the edge of color 3 being h_i). The edge incident with colors 2 and 3 gets color 1; the edges are alternately colored 5, 1, 5, $\cdots$ and the last edge will receive color 5.

It is easy to verify that π is a Kotzíg coloring. □

Theorem 21.5 [106] *Every cubic graph has a Kotzíg coloring.*

Proof. Suppose the Theorem is false; let G be the smallest cubic graph with no Kotzíg coloring. From Lemma 21.4, G must contain no one-factor; therefore, from Exercise 6.3, G contains at least two bridges. Without loss of generality we can assume that G has the form illustrated in Figure 21.2: three subgraphs B_1, A and B_2, together with two edges b_1 and b_2 (bridges), where b_i joins B_i to A; neither B_1 nor B_2 contains a bridge. We write $b_i = u_i v_i$ where u_i lies in B_i.

Construct a new graph G' from G as follows. Delete B_1 and B_2, except for the vertices u_1 and u_2. Then add two new vertices, x and y say, and five new edges xy, xu_1, xu_2, yu_1 and yu_2. G' is shown in Figure 21.3. Since G' is smaller than G, it has a Kotzíg coloring π'. Say u_1v_1 and u_2v_2 receive colors c_1 and c_2 in that coloring.

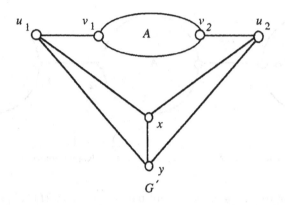

Figure 21.3   A graph used in constructing a Kotzíg coloring.

Next, consider the graphs G_1 and G_2: G_i is constructed from B_i by taking two copies of B_i and adding an edge joining the two copies of u_i. Since G_i contains only one bridge, it has a Kotzíg coloring. By permuting the names of the colors cyclically, if necessary, one obtains a Kotzíg coloring π_i of G_i with the bridge in color c_i.

To construct a Kotzíg coloring π of G, color all edges of B_i with the colors they receive under π_i in one of the copies of B_i in G_i, for $i = 1, 2$. Color all other edges of G with the colors they received under π'. $\square$

We are now in a position to prove

Theorem 21.6 [106] *If G is a cubic graph and H is a cycle of length at least 4, then $G \times H$ has a one-factorization.*

Proof. Say H is of length n. If n is even then the result follows from Corollary 21.1.1. So we can assume n is odd: say $n = 5 + 2r$, where r is a nonnegative integer. Suppose H has vertex-set $\{0, 1, \cdots, n-1\}$ and G has vertex-set $\{a_1, a_2, \cdots, a_g\}$.

Select a Kotzíg coloring π of G. If $a_i a_j$ receives color c under π, then apply color $c + k$ to $(a_i, k)(a_j, k)$ for $k = 1, 2, 3, 4$, and color c to edge $(a_i, k)(a_j, k)$ when $k = 0$ or $k \geq 5$ (all addition is carried out modulo 5). If the two colors missing at vertex a_i under π are x and $x + 1$, then the edges of the cycle

$$(a_i, 0), (a_i, 1), (a_i, 2), \cdots, (a_i, n - 1)$$

are successively colored $x + 1, x + 2, x + 3, x + 4, x, x + 1, x, x + 1, x, \cdots$. The final result is a 5-coloring of $G \times H$. □

Observe that, although this proof is quite satisfactory, it is indirect: the idea of a Kotzíg coloring is needed but the concept does not appear in the statement of the Theorem. Can it be avoided — can a direct proof be found? This and some other open problems are posed in the Exercises.

In view of Theorem 21.6, it is natural to ask about $G \times C_3$, where G is cubic. The graph $N \times C_3$ does not have a one-factorization. In fact, we can prove more.

Theorem 21.7 *If N is the graph of Figure 3.1, then $N \times K_n$ has no one-factorization for n odd.*

Proof. It is convenient to relabel the vertices of N; we assume that the labeling is consistent with that shown in the first part of Figure 23.2, below, and we assume that K_n has vertex-set T. Consider the set $S = \{U, V, W, X, Y\} \times T$. As this set has $5n$ vertices, every one-factor of $N \times K_n$ must have one or more edges with one vertex in S and one vertex outside that set. But $N \times K_n$ contains only n such edges, namely the edges $(Y, i)(Z, i)$ for i in T. So we cannot construct the $n + 2$ disjoint factors required for a one-factorization. □

The argument of Theorem 21.7 can be generalized to the case where G is any cubic graph with a bridge (see Exercise 21.6). In an attempt to clarify the situation, Kotzíg [106] asked: if G is a bridgeless cubic graph, does $G \times C_3$ necessarily have a one-factorization? For example, if P is the Petersen graph, then $P \times K_3$ has a one-factorization [88, 158]. In the next Chapter we prove that the answer to Kotzíg's question is "yes".

Exercises 21

21.1 Prove directly that $P \times K_3$ has a one-factorization.

21.2 Construct a one-factorization of $P \times K_n$ for every odd n, $n \geq 5$.

21.3 (?) Prove that $G \times C_n$ has a one-factorization for every cubic graph G and every odd $n \geq 5$, by a direct method.

21.4 (?) Is it true that, if G is a regular graph of degree $2d + 1$ and $n > 2d + 1$, then $G \times C_n$ has a one-factorization?

21.5 (?) Suppose G is a regular graph of degree $2d + 1$, and H is a critical graph on at least $2d + 2$ vertices. Is $G \times H$ necessarily of class 1?

21.6 Suppose G is a bridgeless cubic graph, and n is any odd positive integer. Prove that $G \times K_n$ has no one-factorization.

KOTZIG'S PROBLEM

To prove that $G \times C_3$ has a one-factorization whenever G is a bridgeless cubic graph, we need some notation, a new definition and several Lemmas.

We need to discuss $C_k \times C_3$. We shall write the vertices of $C_k \times C_3$ as the pairs (i, j), where $i \in \mathbb{Z}_k$ and $j \in \mathbb{Z}_3$, and i_j denotes the edge from (i, j) to $(i+1, j)$. We call i_j the edge *in column j in triangle i*. (Such an edge is always a horizontal edge.) For any a, b and c, we say $(i-1)_a$ *precedes* i_b, and $(i+1)_c$ *succeeds* i_b.

We shall use two standard decompositions of $C_k \times C_3$, one for odd k and one for even k. The examples for $k = 5$ and $k = 6$ are shown in Figures 22.1 and 22.2; other cases can be constructed by repeating the pattern enclosed in a box in the Figure.

Lemma 22.1 *Suppose e, f and g are horizontal edges of $C_k \times C_3$, such that no two of them belong to the same triangle. Then one can decompose $C_k \times C_3$ into two Hamiltonian cycles in such a way that e belongs to one cycle and f and g belong to the other.*

Proof. We consider several cases, depending on the columns in which e, f and g occur.

Case 1. *e, f and g are all in the same column.* If k is odd, we can use the standard decomposition with $e = 1_0$. If k is even, it may be that f or g follows e — without loss of generality, f follows e. If g precedes e, take $e = (k-2)_1$, while if g does not precede e, take $e = (k-1)_2$. If neither f nor g follows e, take $e = (k-2)_2$.

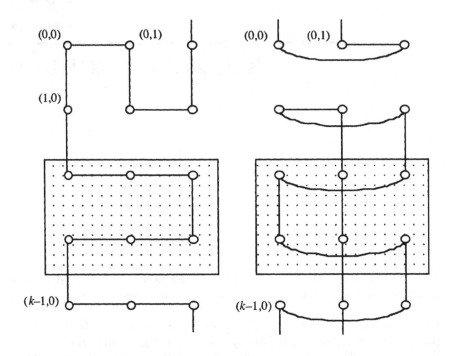

Figure 22.1 Hamiltonian decomposition of $C_k \times C_3 : k$ odd.

Case 2. *e is in one column, f and g in another.* If k is odd, take $e = 0_0$ and take f and g to be any edges in column 2. If k is even, say f precedes and g succeeds e. Then put $e = 1_0$, $f = 0_2$ and $g = 2_2$. Otherwise, if neither f nor g precedes e, put $e = (k-1)_1$ and put f and g in column 2; if f or g precedes e then neither succeeds it, and we may take $e = (k-2)_0$ and again f and g are in column 2.

Case 3. *e and f in one column, g in another.* If k is odd, put $f = 0_1$, e in column 1 and g in column 2. If k is even, we can take $f = (k-1)_2$ if e succeeds f and $f = (k-2)_2$ otherwise. Then e will necessarily be in the right-hand cycle in Figure 22.2 and whichever triangle contains g, it has an edge either in column 0 or in column 1 in the upper cycle; choose such an edge for g.

Case 4. *e, f, g are all in different columns.* If k is odd, set $e = 0_2$ and f and $g = i_0$ and j_1, where $0 < i < j < k$. If k is even, the same assignment works except in the case where it would involve $j = k - 2$. If that occurs — neither

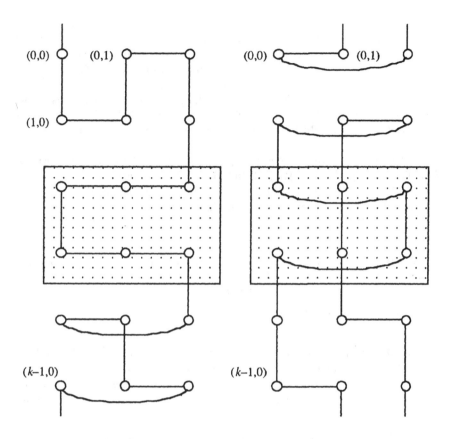

Figure 22.2 Hamiltonian decomposition of $C_k \times C_3$; k even.

f nor g precedes e, but one of them (f say) precedes e by two steps — take $e = 1_2$, $f = (k-1)_1$ and G in column 0. □

The above proof fails when $k = 4$, and in fact the Lemma is false when $k = 4$. (See Exercise 22.2.)

Suppose G is a graph with a spanning subgraph H. Then $G \div H$ denotes the multigraph obtained from G by contracting the edges of H. The vertices will be the components of H. The new graph contains one edge for every edge of G; the edges of H correspond to loops in $G \div H$. It is clear that the edge-connectivity of $G \div H$ is no less than that of G.

We use the same name for an edge of $G \div H$ as for the corresponding edge of G; no confusion will arise. If A is a set of vertices of G, we write $G \div A$ to mean $G \div \langle H \rangle$, where $\langle H \rangle$ is as usual the subgraph induced by A.

In particular, suppose G is a bridgeless cubic graph. Then G contains a one-factor, F say, by Corollary 6.5.1. Write H for the two-factor $G \backslash F$. Then the multigraph $G \div H$ is called the *cycle graph* of G relative to F [168, 169].

The following Theorem of general graph theory was proven independently by Nash-Williams [121] and Tutte [151] in 1961. If $\mathcal{P}$ is any partition of the vertices of a graph or multigraph G, define $z_G(\mathcal{P})$ to be the number of edges of G with endpoints in different parts of $\mathcal{P}$. (The function $z_G(W)$ of Chapter 6 equals $z_G(\{W, V(G) \backslash W\})$.)

Lemma 22.2 *A graph or multigraph G contains k edge-disjoint spanning trees if and only if*

$$z_G(\mathcal{P}) \geq k(|\mathcal{P}| - 1)$$

for every partition $\mathcal{P}$ of $V(G)$. □

Lemma 22.3 *Suppose G is a 3-connected cubic graph other than K_4 and e is some edge of G. There is a decomposition of G into a one-factor F and a two-factor H such that e is in F and $G \div H$ contains edge-disjoint spanning trees R and B, neither of which contains e. Moreover H contains no 4-cycles.*

Proof. We proceed by induction on the size of G. The Lemma is true if G has six vertices (the cubic graphs on six vertices are both Hamiltonian). The induction hypothesis is that the Lemma is true for graphs with fewer vertices than G. Four different cases arise.

(i) Suppose e is a member of a non-trivial 3-edge cut $(X, Y) = \{e, f, g\}$ in G. We consider the multigraphs $G \div X$ and $G \div Y$, in each of which (X, Y) is a trivial 3-cut. Neither multigraph can be K_4. By hypothesis, $G \div Y$ can be factored as $F_X \cup H_X$, where F_X is a one-factor containing e and H_X is a two-factor containing no 4-cycle, such that $(G \div Y) \div H_X$ has edge-disjoint spanning trees R_X and B_X which miss e. $G \div X$ has a similar decomposition $F_Y \cup H_Y$. Then $F = F_X \cup F_Y$ and $H = H_X \cup H_Y$ form a decomposition of G into a one-factor and a two factor (since f and g must be in the same cycle in H_X, and also in the same cycle in H_Y, both will be contained in the same cycle of H, and no problem arises). Then $R_X \cup R_Y$ and $B_X \cup B_Y$ are edge-disjoint spanning trees in $G \div H$. H contains no 4-cycle.

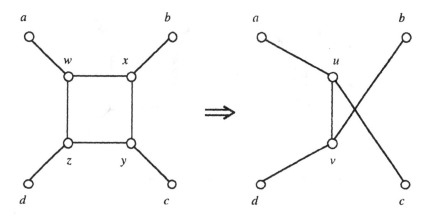

Figure 22.3 Reduction from G to G'.

(ii) Suppose G contains a non-trivial 3-cut $(X,Y) = \{f,g,h\}$, of which e is not a member. Suppose e is in $\langle Y \rangle$. Select a decomposition $G \div Y = F_X \cup H_X$, as guaranteed by the induction hypothesis, with e in F_X. At least one member of (X,Y), say h, will be in F_X, and we may assume that h is not in B_X (exchange the names R_X and B_X if necessary). Then select a decomposition $F_Y \cup H_Y$ of $G \div X$, with the properties guaranteed by the hypothesis, where h is in F_Y, and assume h is not in B_Y. Now proceed as in case (i).

(iii) Suppose G contains no non-trivial 3-cut, but G contains a 4-cycle which does not contain e. Suppose the 4-cycle is (w,x,y,z), and the other four edges of G incident with it are (w,a), (x,b), (y,c) and (z,d), as shown on the left in Figure 22.3. (The 4-cycle can contain no chord or multiple edge: if (w,y) were a chord or (w,x) a double edge then either (d,z) would be a bridge, whence G is not 2-connected, or else G has only four vertices.)

Delete the cycle and the four edges incident with it, and replace them with two new vertices u and v and the edges (u,a), (u,c), (v,b), (v,d) and (u,v), as shown in Figure 22.3. Call the new graph formed G'. Then G' is smaller than G, so it can be decomposed into a one-factor F' and a two-factor H' which satisfy the Lemma, with e in F'.

We show how to expand this decomposition to a suitable decomposition $G = F \cup H$. First, suppose (u,v) is in H. Then we replace (u,v) by three edges which pass through all of $\{w,x,y,z\}$. For example, the segment $\cdots a,u,v,b\cdots$

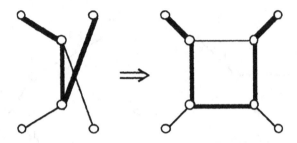

Figure 22.4 Expansion from G' to G when $(u,v) \in H'$.

is replaced by $\cdots a, w, z, y, x, b \cdots$ (see Figure 22.4), $\cdots a, u, v, d \cdots$ is replaced by $\cdots a, w, x, y, z, b \cdots$, and so on. In this case $G \div H$ will be isomorphic to $G' \div H'$, so it contains two edge-disjoint trees, missing e if necessary.

If (u, v) is not in H' then the situation is shown in Figure 22.5, where the grey line represents a path (not necessarily one edge) in H'. If the expansion shown in that Figure is made, then the resulting H has one cycle containing w, x, y and z, and $G \div H$ is isomorphic to $G' \div H' \div \{(u,v)\}$. The two edge-disjoint spanning trees in $G' \div H'$ become two edge-disjoint spanning trees in $G \div H$, neither of which contains e.

In either case H contains no 4-cycle, because H' contains none.

(iv) Finally, suppose G contains no non-trivial 3-cut, and at most one 4-cycle, and if it contains a 4-cycle then that cycle contains e. From Lemma 6.5.1, G contains a one-factor F which includes e. We write H for the 2-factor $G \setminus F$.

Consider any partition $\mathcal{P} = \{P_1, P_2, \cdots, P_r\}$ of the vertices of $G \div H$. If there were only three edges joining P_i to the rest of $G \div H$, then they would correspond to a 3-cut in G. Of the edges incident with a vertex of G, two must be members of H, so at most one is an edge of $G \div H$. So no two members of the 3-cut are adjacent to the same vertex of G, and the cut is non-trivial. This is impossible. Therefore each P_i is incident with at least four edges joining it to other parts of $\mathcal{P}$, and $z_G(\mathcal{P}) \geq 2r$. If we exclude the edge e from the edge-set, then $z_G(\mathcal{P}) \geq 2r - 1 > 2(r - 1)$. By Lemma 22.2, $G \div H$ contains two edge-disjoint spanning trees, both missing e.

Since G contains no 4-cycle except possibly one which includes e, H can contain no 4-cycle.   □

Lemma 22.4 *Suppose the graph or multigraph G is the union of two disjoint acyclic graphs R and B. Then the vertices of G can be ordered v_1, v_2, $\cdots$, v_n, where v_1 is arbitrary, in such a way that v_i is incident with at most three edges joining it to earlier vertices in the sequence, of which at most two are edges from the same member of $\{R, B\}$.*

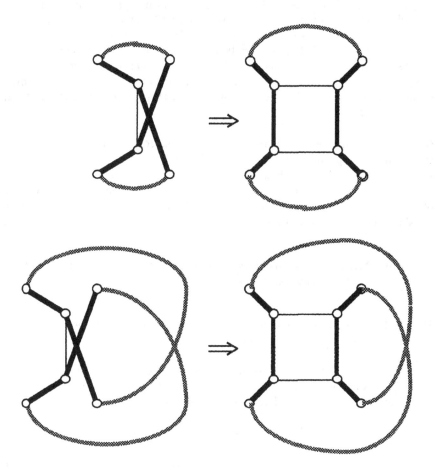

Figure 22.5 Expansion from G' to G when $(u,v) \notin H'$.

Proof. We proceed by induction on the number n of vertices of G. The proposition is obviously true for $n = 1$. Suppose G satisfies the conditions of the Lemma and has n vertices; let v be some vertex of G.

R may be a spanning tree in G. If not, augment the vertex-set of R to make it span G and add new edges joining components until a spanning tree R' is formed. Similarly, embed B in a spanning tree B'. Then $R' \cup B'$ has $2n - 2$ edges, so the sum of its degrees is $4n - 4$. Moreover, each vertex has degree at least 2. So $R' \cup B'$ has at least two vertices of degree less than 4. Select a vertex v_n, $v_n \neq v$, whose degree is less than 4. Then v_n is incident with at least one edge of R' and at least one edge of B', so it lies on less than three edges of either tree. So, in G, v_n lies on at most three edges, at most two from either R or B.

$G - v_n$ is the union of the disjoint acyclic graphs $R - v_n$ and $B - v_n$. So, by the induction hypothesis, we may order the vertices of $G - v_n$ as $v_1, v_2, \cdots,$ v_{n-1}, where $v_1 = v$ and v_i belongs to at most three edges joining it to earlier vertices in the sequence, at most two of which are edges from the same member of $\{R - v_n, B - v_n\}$. If we append v_n to the end of this sequence, we have satisfied the requirements of the Lemma.   □

Theorem 22.5 *If G is a 3-connected cubic graph then $G \times C_3$ can be factored into two Hamilton cycles and a one-factor.*

Proof. It is easy to see that $K_4 \times C_3$ can be decomposed in the desired way (see Exercise 22.4), so we assume G is not K_4.

We suppose $G = F \cup H$ is a decomposition of the kind guaranteed by Lemma 22.3. F is a one-factor, H is a two-factor such that $G \div H$ contains two disjoint spanning trees R and B, and H contains no 4-cycle.

Using Lemma 22.4, order the vertices of $G \div H$ as $v_1, v_2, \cdots,$ where each vertex lies on at most three edges of R and B which join it to earlier vertices in the sequence, at most two from the same tree. Write V_i for the cycle of H corresponding to v_i. None of the V_i is a 4-cycle.

The edges of $G \times C_3$ will be colored in red, blue and white, so that the red and blue sets each form Hamilton cycles, and the white edges form a one-factor. Initially the vertical edges derived from edges of F are all colored white (this is already a one-factor, but some recoloring may be necessary). The remaining edges are colored in the following order: first all of $V_1 \times C_3$, then all of $V_2 \times C_3$, and so on. In each case $V_i \times C_3$ will be decomposed into two Hamilton cycles, using Lemma 22.3; one will be colored red and the other blue.

Say $V_h \times C_3$ has been colored for all $h < i$. Suppose there is an edge of R from v_i to v_j in $G \div H$, where $j < i$; say it is (x, y) (where $x \in V_j$ and $y \in V_i$). Since $V_j \times C_3$ has been decomposed into red and blue Hamilton cycles, triangle x of $V_j \times C_3$ has at least one red edge; select one, say edge x_a. Then edge y_a in $V_i \times C_3$ is required to be red. Similarly, for each edge of B joining v_i to an earlier vertex v_k in $G \div H$, we select a blue edge in $V_k \times C_3$ and require the corresponding edge in $V_i \times C_3$ to be blue. In this way we impose colors on at most three edges of $V_i \times C_3$; if there are three edges, the same color is not imposed on them all; and no two of the edges are in the same horizontal triangle in $V_i \times C_3$ (if two were in triangle y, then vertex y of G would lie on two edges in a cycle and also two edges corresponding to edges of B and H, so y would have degree at least 4, which is impossible since G is cubic). So, by Lemma 22.1, we can choose a suitable decomposition of $V_i \times C_3$ into two Hamilton cycles.

Finally we carry out some recoloring so that the red edges form a Hamilton cycle. To each edge (v_j, v_i) of R, there corresponds an edge (x, y) of F (where x is a vertex of V_j and y is in V_i). There will exist two red edges in $V_j \times C_3$ and $V_i \times C_3$ of the form x_a and y_a; the edges $(x, a)(y, a)$ and $(x, a+1)(y, a+1)$ are white. Exchange the colors so that these two edges become red and x_a and y_a become white. The white edges still form a one-factor, and if the exchange is carried out for all edges of R then the red edges will form a Hamilton cycle. A similar process is carried out for the edges of B, resulting in the required decomposition. □

Theorem 22.6 *If G is a bridgeless cubic multigraph, then $G \times C_3$ has a one-factorization.*

Proof. We proceed by induction on the number of 2-cuts in G. If there are no 2-cuts, then Theorem 22.5 applies. So we assume G contains n 2-cuts, and assume the Theorem is true of multigraphs containing fewer than n 2-cuts. No 2-cut in G can be trivial, because if it were then the third edge touching the common vertex would be a bridge. Select a cut $\{ab, cd\}$ such that $G - \{ab, cd\}$ is the disjoint union of graphs G_1 and G_2, a and c being vertices of G_1 and b and d being vertices of G_2, where $H_1 = G_1 + ac$ contains fewer than n 2-cuts and $H_2 = G_2 + bd$ contains no 2-cuts. (See Figure 22.6.)

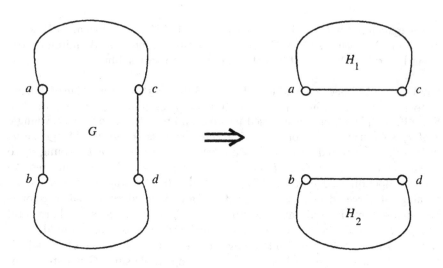

Figure 22.6   Construction for Theorem 22.6.

By the induction hypothesis, $H_1 \times C_3$ has a one-factorization. Select such a factorization. If the three edges $a_0 c_0$, $a_1 c_1$ and $a_2 c_2$ are in different factors, color all the edges blue in the factors containing $a_0 c_0$ and $a_1 c_1$, and color all the edges red in the factor containing $a_2 c_2$ and in one other factor. If the three edges together belong to two factors, color all edges in that factor red; if they are all in the same factor, color that factor and some other factor red. In both these cases, color red all the edges in two further factors. In every case, the edges in the remaining factor are colored white.

We now use Theorem 22.5 to decompose $H_2 \times C_3$ into two Hamilton cycles, colored red and blue, and a one-factor. (H_2 cannot contain multiple edges — see Exercise 22.5 — so the Theorem may be applied.) In particular, when we decompose H_2 into a one-factor and a two-factor and a two-factor we specify that the new edge bd is not in the one-factor (select another edge through b to be in the one-factor) and we choose the cycle through bd to be V_1, because the decomposition of $V_1 \times C_3$ can be chosen without restriction. Select that decomposition so that $b_i d_i$ is the same color as $a_i c_i$, for every i. The recoloring process does not affect the vertical edges in the $V_i \times C_3$.

Finally, color $a_i b_i$ and $c_i d_i$ in the same color as $a_i c_i$ and $b_i d_i$, for every i, and color the other edges of $G \times C_3$ in the same way as in $H_1 \times C_3$ and $H_2 \times C_3$. Clearly the set of red edges is a union of even cycles, so it decomposes into two one-factors, and similarly for the blue edges, while the white edges form a one-factor. □

Exercises 22

22.1 Prove that $C_3 \times C_3$ has precisely two different one-factorizations, up to isomorphism.

22.2 Prove that, in the notation of Lemma 22.1, $C_4 \times C_3$ has no factorization into two Hamilton cycles in which 0_0 and 2_0 belong to one cycle and 1_0 belongs to the other. However, the following weaker Lemma will be sufficient for our purposes.

22.3 Suppose e and f are any two horizontal edges in $C_4 \times C_3$. Prove that there is a decomposition of $C_4 \times C_3$ into Hamilton cycles such that e and f are in different cycles.

22.4 Find an edge-disjoint decomposition of $K_4 \times C_3$ into two Hamilton cycles and a one-factor.

22.5 Prove that if a connected cubic multigraph contains a multiple edge xy then there is a two-edge cut adjacent to xy.

22.6 (?) Is there a bridgeless cubic graph which cannot be decomposed into two Hamilton cycles and a one-factor?

OTHER PRODUCTS

The *tensor product* $G \otimes H$ of graphs G and H is defined to have as its vertices the ordered pairs (a, b) where a is a vertex of G and b is a vertex of H; (a_1, b_1) is adjacent to (a_2, b_2) if and only if $a_1 a_2$ is an edge of G and also $b_1 b_2$ is an edge of H.

The *wreath product* or *composition* of two graphs G and H is defined as follows: if G is a graph with vertices $p_1, p_2, \cdots, p_g$, and H is any graph, then the wreath product $G[H]$ consists of the disjoint union of g copies $H_1, H_2, \cdots, H_g$ of H, to which are added all the edges joining vertices in H_x to vertices of H_y if and only if p_x is adjacent to p_y in G. In other words, one has a copy of G; each vertex is replaced by a copy of H, and each edge is replaced by a copy of $K_{h,h}$, where h is the number of vertices of H. Observe that $G[H]$ and $H[G]$ need not be isomorphic.

For comparison, Figure 23.1 shows $P_3 \times K_3$, $P_3 \otimes K_3$ and $P_3[K_3]$.

If $G = G_1 \cup G_2$ is a factorization, then

$$G[H] = G_1[H] \oplus G_2[\overline{K}]. \tag{23.1}$$

Similarly, if G_1 and G_2 are vertex-disjoint, then $G_1[H]$ and $G_2[H]$ are vertex-disjoint. Another important property is

$$(G[H])[J] = G[H[J]]. \tag{23.2}$$

The notations $G\{m\}$ and $G(m)$ are used for $G \otimes K_m$ and $G[\overline{K}_m]$ respectively. In this notation

$$G(m) = G\{m\} \oplus mG, \tag{23.3}$$

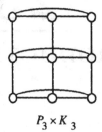
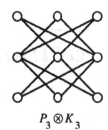
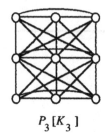

$P_3 \times K_3$ $P_3 \otimes K_3$ $P_3[K_3]$

Figure 23.1 Products of P_3 with K_3.

and

$$G[H] = (G \times H) \oplus G\{v(H)\}. \tag{23.4}$$

Suppose G is the sum of a series of graphs

$$G = G_1 \oplus G_2 \oplus \cdots \oplus G_k.$$

Then

$$G\{m\} = G_1\{m\} \oplus G_2\{m\} \oplus \cdots \oplus G_k\{m\}, \tag{23.5}$$
$$G \otimes H = (G_1 \otimes H) \oplus (G_2 \otimes H) \oplus \cdots \oplus (G_k \otimes H). \tag{23.6}$$

Moreover, if each G_i has a one-factorization, then G will have a one-factorization.

Lemma 23.1 [128] *If G is a regular graph, then $G\{2m\}$ has a one-factorization.*

Proof. Select a one-factorization $\mathcal{F} = F_1, F_2, \cdots, F_{2m-1}$ of K_{2m}. Then

$$K_{2m} = F_1 \oplus F_2 \oplus \cdots \oplus F_{2m-1},$$

so, using (23.6),

$$G\{2m\} = G \otimes K_{2m}$$
$$= (G \otimes F_1) \oplus (G \otimes F_2) \oplus \cdots \oplus (G \otimes F_{2m-1}),$$

and it suffices to show that each $G \otimes F_i$ has a one-factorization. But F_i is the union of m disjoint edges, so $G \otimes F_i$ is isomorphic to $mG\{2\}$. But $G\{2\}$ is bipartite, and it is regular because G is regular, so it has a one-factorization (see Corollary 8.3.1). So $mG\{2\}$ also has a one-factorization. □

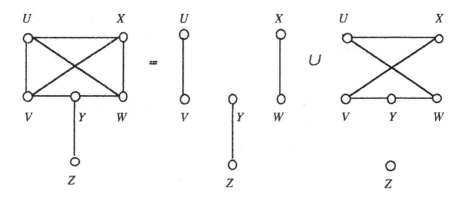

Figure 23.2   Factorization of the subgraph N_1.

Theorem 23.2 [128, 91]   *The wreath product $G[H]$ has a one-factorization in the following cases:*

(i) *both G and H contain one-factors;*

(ii) *G is non-null and has a one-factorization;*

(iii) *H is non-null and has a one-factorization;*

(iv) *H is cubic and G is an odd cycle of length greater than 3;*

(v) *H is a bridgeless cubic graph and $G = C_3$.*

Proof. In each case we use the decomposition

$$G[H] = (G \times H) \oplus G\{v(H)\}.$$

and observe that each summand has a one-factorization. The factorizations of $G \times H$ are supplied by Corollary 21.1.1 and Theorems 21.3, 21.6 and 22.6, while the factorizations of the graphs $G\{v(H)\}$ come from Lemma 23.1.   □

The sufficient conditions in the above Theorem are not necessary. We prove this using the graph N of Figure 3.1.

Theorem 23.3 $N[\overline{K}_3]$ *has a one-factorization.*

Proof. We can consider N as the union of three graphs, N_1, N_2 and N_3, where N_1 is shown on the left of Figure 23.2, N_2 is formed by rotating through

120° around the common central vertex Z, and N_3 by rotating through a further 120°. In converting N to $N[\overline{K}_3]$, we assume that vertex A becomes $\{A_1, A_2, A_3\}$. Clearly

$$N[\overline{K}_3] = N_1[\overline{K}_3] \cup N_2[\overline{K}_3] \cup N_3[\overline{K}_3].$$

Let $N_i(j)$ denote $N_i[\overline{K}_3]$ with all of Z_1, Z_2, Z_3 deleted except for Z_j. We shall exhibit an edge-disjoint decomposition

$$N_i[\overline{K}_3] = N_{i1} \cup N_{i2} \cup N_{i3}$$

where N_{ij} is the union of three disjoint one-factors of $N_i(j)$. Then it follows that each of the following is a set of three one-factors of $N[\overline{K}_3]$:

$$N_{11} \cup N_{22} \cup N_{33}$$
$$N_{12} \cup N_{23} \cup N_{31}$$
$$N_{13} \cup N_{21} \cup N_{32}$$

Together they form a one-factorization of $N[\overline{K}_3]$.

Here are the components of $N_1[\overline{K}_3]$. In each case, each row is a one-factor.

$$N_{11} : \begin{cases} U_1V_2 & U_2Y_2 & U_3V_1 & V_3X_3 & X_1W_3 & X_2W_1 & Y_1W_2 & Y_3Z_1 \\ U_1V_3 & U_2V_1 & U_3Y_1 & V_2X_2 & X_1W_2 & X_3W_1 & Y_2Z_1 & Y_3W_3 \\ U_1Y_3 & U_2V_2 & U_3V_3 & V_1X_1 & X_2W_2 & X_3W_3 & Y_1Z_1 & Y_2W_1 \end{cases}$$

$$N_{12} : \begin{cases} U_1X_2 & U_2X_3 & U_3Y_2 & V_1W_3 & V_2X_1 & V_3W_2 & Y_1W_1 & Y_3Z_2 \\ U_1Y_1 & U_2X_1 & U_2X_2 & V_1X_3 & V_2W_3 & V_3W_1 & Y_2Z_2 & Y_3W_2 \\ U_1X_1 & U_2Y_3 & U_3X_3 & V_1W_1 & V_2W_2 & V_3X_2 & Y_1Z_2 & Y_2W_3 \end{cases}$$

$$N_{13} : \begin{cases} U_1Y_2 & U_2V_3 & U_3X_1 & V_1X_2 & V_2W_1 & X_3W_2 & Y_1W_3 & Y_3Z_3 \\ U_1X_3 & U_2Y_1 & U_3V_2 & V_1W_2 & V_3X_1 & X_2W_3 & Y_2Z_3 & Y_3W_1 \\ U_1V_1 & U_2X_2 & U_3Y_3 & V_2X_3 & V_3W_3 & X_1W_1 & Y_1Z_3 & Y_2W_2 \end{cases}$$

The components of $N_2[\overline{K}_3]$ and $N_3[\overline{K}_3]$ are formed by rotation.   □

It is appropriate to exhibit one example of a wreath product with no one-factorization. We use $N[K_3]$; clearly other cubic graphs without one-factors could be used instead of N, and the method can be generalized to regular graphs of higher degree without one-factors.

Theorem 23.4 $N[K_3]$ *has no one-factorization.*

Proof. $N[K_3]$ is a regular graph of degree 11, so any one-factorization contains 11 factors. If we denote the vertices as we did above, then $S = \{U_1, U_2, U_3, V_1, \cdots, Y_3\}$ contains 15 vertices. Since 15 is odd, each one-factor of $N[K_3]$ contains at least one edge with one endpoint in S and one outside. This edge must be of the form $Y_i Z_j$. Only nine such edges exist, so eleven factors are impossible. □

This argument is based on the fact that $N[K_n]$ has degree $4n - 1$, and that the number of edges $Y_i Z_j$ is n^2; so, for n odd, a factorization is impossible when $n^2 < 4n - 1$. But the only case ruled out is $n = 3$. It is quite possible that $N[K_5]$ has a one-factorization.

We now look briefly at tensor products.

Theorem 23.5 [128] *If G and H are regular graphs and at least one of them has a one-factorization then $G \otimes H$ has a one-factorization.*

Proof. Say H has a one-factorization $\{F_1, F_2, \cdots, F_k,$ so that

$$G = F_1 \oplus F_2 \oplus \cdots \oplus F_k,.$$
$$G \otimes H = (G \otimes F_1) \oplus (G \otimes F_2) \oplus \cdots \oplus (G \otimes F_k).$$

Suppose G has $2n$ vertices. Then each F_i is isomorphic to nK_2. So $G \otimes F_i$ is isomorphic to $n(G \otimes K_2) = nG\{2\}$. From Lemma 23.1, $G\{2\}$ has a one-factorization, and so also do $G \otimes F_i$ and $G \otimes H$. □

Several other results on factors and factorizations of tensor products have been discovered; most require some familiarity with group theory. For example, Alspach and George [3] have proven the following result.

Theorem 23.6 *If G is a regular graph of degree n, where n is a prime power, then $G \otimes K_n$ has a one-factorization.* □

As a consequence, $N \otimes K_3$ has a one-factorization, in contrast with the wreath product case.

Several related results on numbers of edge-disjoint one-factors in tensor products are obtained, from largely algebraic considerations, in their paper and in George's dissertation [75].

Exercises 23

23.1 (?) What can be said about one-factorizations of $G[C_n]$ when G is regular and n is odd?

23.2 (?) Does $N[K_5]$ have a one-factorization?

23.3 Suppose G and H are connected regular graphs, both $v(G)$ and $\Delta(H)$ are odd, $\Delta(H) < v(G)$, and H has a bridge xy.

   (i) Prove that each component of $H - xy$ has an odd number of vertices.

   (ii) Hence prove that $G \otimes H$ has no one-factorization. [75]

23.4 The *strong tensor product* $G \odot H$ is defined to have vertex-set $V(G) \times V(H)$, and (a_1, b_1) is adjacent to (a_2, b_2) if and only if a_1 is adjacent to a_2 and also b_1 is adjacent to or equal to b_2.

   (i) Prove that

$$G \odot H = (G \otimes H) \oplus \bigcup_{i=1}^{m}(G \times \{v_i\}),$$

   where

$$V(H) = \{v_1, v_2, \cdots, v_m\}.$$

   (ii) Are $G \odot H$ and $H \odot G$ isomorphic in general?

   (iii) Suppose G has a one-factorization and H is regular. Prove that $G \odot H$ has a one-factorization. [128]

23.5 The *symmetric difference product* $G \vee H$ is defined to have vertex-set $V(G) \times V(H)$, and (a_1, b_1) is adjacent to (a_2, b_2) if and only if either a_1 is adjacent to a_2 in G or b_1 is adjacent to b_2 in H,.but not both.

   (i) Prove that

$$G \vee H = (G \times H) \oplus (\overline{G} \otimes H) \oplus (G \otimes \overline{H}).$$

   (ii) Suppose either G or H is a complete graph on an even number of vertices. Prove that $G \vee H$ has a one-factorization. [75]

23.6 The *normal product* $G \circ H$ is defined to have vertex-set $V(G) \times V(H)$, and (a_1, b_1) is adjacent to (a_2, b_2) if and only if one of:

$$(a_1 \sim a_2 \text{ or } a_1 = a_2) \quad \text{and} \quad b_1 \sim b_2;$$
$$(b_1 \sim b_2 \text{ or } b_1 = b_2) \quad \text{and} \quad a_1 \sim a_2$$

   is true.

(i) Prove that
$$G \circ H = (G \times H) \oplus (G \otimes H).$$

(ii) Are $G \circ H$ and $H \circ G$ isomorphic in general?

(iii) Suppose G and H have one-factorizations, prove that $G \circ H$ has a one-factorization also.

ONE-FACTORIZATIONS
OF K_{10}

The factors are labeled $\mathcal{F}1$ through $\mathcal{F}396$, in the order used in [73]. Every factorization contains the factor 01 23 45 67 89. In listing the other eight factors, a word such as 514283769 denotes the one-factor 05 14 28 37 69.

213465879	213465879	213465879	213465879	213465879	213465879
312495768	312495768	312475968	312475968	312495768	312475968
415263978	415263978	415263978	415263978	415263978	415263978
514283769	514283769	514283769	514283769	514283769	514283769
617243859	617243859	617253849	617253849	617293548	619273548
716293548	716293548	716293548	716293548	716243859	716253849
819253647	819273456	819243657	819273456	819273456	817293456
918273456	918253647	918273456	918243657	918253647	918243657
$\mathcal{F}1$	$\mathcal{F}2$	$\mathcal{F}3$	$\mathcal{F}4$	$\mathcal{F}5$	$\mathcal{F}6$
213465879	213465879	213465879	213465879	213465879	213465879
312475968	312495768	312495768	312475968	312495768	312495768
415263978	415263978	415263978	416283957	416273859	415283769
514283769	516293847	516293748	518263749	518293647	514263978
617253849	619253748	617243859	617293548	614253978	618293547
718293456	714283659	714283569	715243869	719283456	719243856
819243657	817243569	819273456	819273456	815243769	816273459
916273548	918273456	918253647	914253678	917263548	917253648
$\mathcal{F}7$	$\mathcal{F}8$	$\mathcal{F}9$	$\mathcal{F}10$	$\mathcal{F}11$	$\mathcal{F}12$
213465879	213465879	213465879	213465879	213465879	213465879
312495768	312475968	312475968	312475968	312475968	312495768
415283769	415263978	415263978	415263978	415263978	415263978
516293478	516283749	516283749	516283749	518273469	516293847
614273859	614293857	617293548	619273548	617293548	614283759
719263548	719253648	714253869	714293856	716253849	719253648
817243956	817243569	819243657	817253469	819243756	817243569
918253647	918273456	918273456	918243657	914283657	918273456
$\mathcal{F}13$	$\mathcal{F}14$	$\mathcal{F}15$	$\mathcal{F}16$	$\mathcal{F}17$	$\mathcal{F}18$

213465879	213465879	213465879	214365879	213465879	213465879
312495768	312475968	312475968	315264978	312475968	312475968
415263978	415263978	415263978	412375968	415263978	416253978
514283769	516293748	514283769	513274869	514283769	518243769
617293548	617253849	618293457	618293457	619273548	617293548
718243659	714283569	719253648	719243856	718253649	715263849
819273456	819243657	816273549	817253946	817293456	819273456
916253847	918273456	917243856	916283547	916243857	914283657
$\mathcal{F}19$	$\mathcal{F}20$	$\mathcal{F}21$	$\mathcal{F}22$	$\mathcal{F}23$	$\mathcal{F}24$
214365879	213465879	213465879	213465879	213465879	213465879
315264978	312475968	312475968	314256978	312475968	312475968
412375968	415263978	415263978	412395768	415263978	416253978
513284769	514273869	518273649	518273649	516283749	518273469
619243857	617283549	617293548	615293847	617293548	617283549
718293456	718293456	714253869	716283459	714253869	714293856
817253946	819243657	819243756	819243756	819273456	819243657
916273548	916253748	916283457	917263548	918243657	915263748
$\mathcal{F}25$	$\mathcal{F}26$	$\mathcal{F}27$	$\mathcal{F}28$	$\mathcal{F}29$	$\mathcal{F}30$
213465879	213465879	214365879	213465879	213465879	213465879
314256978	314265978	315264978	312495768	312475968	312495768
412385769	412385769	412375968	415263978	416253978	416253978
516293847	516283947	513274869	517243869	518273469	517283469
617283459	619253748	618243957	618293547	617293548	618243759
718263549	718293456	719253846	719283456	715263849	719263548
819243756	815273649	816293547	814273659	819243756	815293647
915273648	917243568	917283456	916253748	914283657	914273856
$\mathcal{F}31$	$\mathcal{F}32$	$\mathcal{F}33$	$\mathcal{F}34$	$\mathcal{F}35$	$\mathcal{F}36$
213465879	213465879	213465879	213465879	213465879	213465879
312495768	312495768	314256978	312495768	312475968	312475968
415263978	415263978	412375968	415263978	415263978	415263978
516293748	517283469	516283947	516293847	516283749	516293748
619283547	614273859	619243857	614283759	614293857	617283549
718253469	719253648	718293456	718243569	718243569	718253469
817243659	816293547	815273649	819273456	819273456	819243657
914273856	918243756	917263548	917253648	917253648	914273856
$\mathcal{F}37$	$\mathcal{F}38$	$\mathcal{F}39$	$\mathcal{F}40$	$\mathcal{F}41$	$\mathcal{F}42$
214365879	213465879	213465879	213465879	213465879	213465879
315264978	312495768	314256978	314256978	312475968	312475968
412375968	415263978	412375968	415273968	415263978	415263978
513284769	517243869	516283947	516283749	518273469	518243769
618293457	614283759	617293548	617293548	617283549	617293548
719243856	716293548	719243856	718263459	714293856	716253849
817253946	819273456	815273649	819243657	819243657	819273456
916273548	918253647	918263457	912384756	916253748	914283657
$\mathcal{F}43$	$\mathcal{F}44$	$\mathcal{F}45$	$\mathcal{F}46$	$\mathcal{F}47$	$\mathcal{F}48$

214365879	213465879	213465879	213465879	213465879	213465879
315264978	314256978	312475968	312475968	314265978	314256978
412375968	412375968	416253978	415263978	415273968	415273968
513274869	516293847	518243769	514283769	516293748	516293748
618243957	618243957	617293548	618273549	617283549	619283457
719283456	719283456	715283649	719253648	719243856	712384956
817293546	817263549	819273456	817293456	812345769	817243659
916253847	915273648	914263857	916243857	918253647	918263547
$\mathcal{F}49$	$\mathcal{F}50$	$\mathcal{F}51$	$\mathcal{F}52$	$\mathcal{F}53$	$\mathcal{F}54$

213465879	213465879	213465879	213465879	213465879	213465879
312475968	314256978	314256978	312475968	314256978	314256978
416293857	412375968	412375968	415263978	412375968	412395768
519273648	516283947	517263948	514283769	518273649	516283749
615243978	619273548	615293847	618293457	615293847	617243859
718263549	715263849	719283456	719243856	719263548	719263548
814253769	817293456	816273549	816273549	817243956	815293647
917283456	918243657	918243657	917253648	916283457	918273456
$\mathcal{F}55$	$\mathcal{F}56$	$\mathcal{F}57$	$\mathcal{F}58$	$\mathcal{F}59$	$\mathcal{F}60$

213465879	213465879	213465879	213465879	214365879	214365879
312475968	312475968	312475968	312475968	315264978	315264978
415263978	415263978	415263978	415263978	412395768	412395768
519273648	516293748	516273849	517283649	517283469	518243769
614293857	617253849	617293548	614293857	619253847	619283547
716283549	718243569	718253469	718243569	718293546	713294856
817253469	819273456	819243756	819273456	816243759	816273459
918243756	914283657	914283657	916253748	913274856	917253846
$\mathcal{F}61$	$\mathcal{F}62$	$\mathcal{F}63$	$\mathcal{F}64$	$\mathcal{F}65$	$\mathcal{F}66$

213465879	213465879	213465879	213465879	213465879	214365879
312475968	314256978	312475968	314256978	312495768	315264978
415263978	415273968	415263978	415273968	415263978	412385769
519273648	516293748	516273849	516283749	516293748	519243768
617253849	617283549	617293548	617293548	619283547	618273459
714283569	718243659	718253469	718243659	718243659	713294856
816293457	819263457	819243657	819263457	817253469	816253947
918243756	912384756	914283756	912384756	914273856	917283546
$\mathcal{F}67$	$\mathcal{F}68$	$\mathcal{F}69$	$\mathcal{F}70$	$\mathcal{F}71$	$\mathcal{F}72$

213465879	213465879	213465879	213465879	213465879	213465879
312495768	314256978	314256978	314256978	314256978	314256978
415263978	412375968	412375968	412375968	417283956	412395768
517293648	516283947	518273649	516273849	518263749	516293847
619283547	618293457	615293847	615283947	619243857	618243759
714253869	719243856	719263548	718293456	715293468	719283456
816273459	815273649	816243957	819243657	812364759	815273649
918243756	917263548	917283456	917263548	916273548	917263548
$\mathcal{F}73$	$\mathcal{F}74$	$\mathcal{F}75$	$\mathcal{F}76$	$\mathcal{F}77$	$\mathcal{F}78$

214365879	213465879	213465879	213465879	213465879	213465879
315264978	314256978	312475968	314256978	314256978	312495768
412375968	412375968	415263978	412395768	412395768	415283769
513284769	518263947	516273849	518263749	518293647	516243978
618243957	619243857	618293457	619273548	617243859	614273859
719253846	716293548	719253648	716243859	719283456	718293456
817293456	815273649	817243569	817293456	816273549	819263547
916273548	917283456	914283756	915283647	915263748	917253648
$\mathcal{F}79$	$\mathcal{F}80$	$\mathcal{F}81$	$\mathcal{F}82$	$\mathcal{F}83$	$\mathcal{F}84$

213465879	213465879	213465879	213465879	213465879	213465879
314256978	314256978	312475968	312475968	314256978	314256978
412395768	412395768	415263978	416253978	412395768	416273859
516293847	519283647	516283749	518273469	519273648	518293647
618243759	617243859	619273548	617293548	618243759	617283549
719263548	718293456	714253869	715283649	716283549	715243968
815273649	816273549	817293456	819243756	817293456	819263457
917283456	915263748	918243657	914263857	915263847	912374856
$\mathcal{F}85$	$\mathcal{F}86$	$\mathcal{F}87$	$\mathcal{F}88$	$\mathcal{F}89$	$\mathcal{F}90$

214365879	213465879	213465879	214365879	213465879	213465879
315264978	314256978	314256978	315264978	312475968	312475968
412395768	415273968	415273968	412395768	416253978	415263978
513284769	519263748	516293847	517243869	518243769	519273648
618243759	612384759	617283459	619253748	617283549	617283549
719253846	716283549	719263548	718293456	714293856	714253869
817293456	817293456	812364957	813274659	819263457	816293457
916273548	918243657	918243756	916283547	915273648	918243756
$\mathcal{F}91$	$\mathcal{F}92$	$\mathcal{F}93$	$\mathcal{F}94$	$\mathcal{F}95$	$\mathcal{F}96$

213465879	213465879	213465879	213465879	213465879	213465879
314256978	312475968	312495768	312475968	314256978	314295678
415273968	415263978	415263978	415263978	415273968	417253869
516283749	514283769	516293847	516293748	516293748	518273649
617283459	617293548	617283459	617283549	617283549	619283547
719263548	718253649	719253648	714253869	718263459	715263948
812374956	819273456	814273569	819273456	819243657	816243759
918243657	916243857	918243756	918243657	912384756	912345768
$\mathcal{F}97$	$\mathcal{F}98$	$\mathcal{F}99$	$\mathcal{F}100$	$\mathcal{F}101$	$\mathcal{F}102$

214365879	214365879	213465879	213465879	213465879	213465879
315284769	315284769	314256978	312475968	314256978	314256978
419253768	417293568	412395768	415263978	412375968	412395768
516243978	519263478	519263748	518243769	516283947	518263749
618293457	618243957	618273549	617283549	618293457	617243859
712384659	712384956	716283459	719253648	719243856	719283456
817263549	813274659	815293647	816293457	817263549	815293647
913274856	916253748	917243856	914273856	915273648	916273548
$\mathcal{F}103$	$\mathcal{F}104$	$\mathcal{F}105$	$\mathcal{F}106$	$\mathcal{F}107$	$\mathcal{F}108$

214365879	214365879	214365879	214365879	213465879	213465879
315284769	315264978	315284769	315274968	312475968	314256978
417293568	412395768	416293578	412385769	416253978	415293768
518273946	518273469	519243768	516293748	518273469	518263947
612345978	619283547	612384957	619283547	617293548	617283459
719243856	713294856	718263459	718243956	719243856	719243856
816253749	816243759	817253946	817263459	815263749	816273549
913264857	917253846	913274856	913254678	914283657	912364857
$\mathcal{F}109$	$\mathcal{F}110$	$\mathcal{F}111$	$\mathcal{F}112$	$\mathcal{F}113$	$\mathcal{F}114$

213465879	213465879	213465879	213465879	213465879	213465879
312475968	314256978	312475968	314275968	314256978	314265978
416283957	412395768	416283957	415283769	412395768	412385769
518273649	517283649	518273649	516243978	518273649	517243968
615293478	618243759	615293478	617253849	617293548	615293748
719263548	716293548	719243856	718293456	719283456	718253649
814253769	819273456	814253769	819263547	816243759	819273456
917243856	915263847	917263548	912364857	915263847	916283547
$\mathcal{F}115$	$\mathcal{F}116$	$\mathcal{F}117$	$\mathcal{F}118$	$\mathcal{F}119$	$\mathcal{F}120$

214365879	213465879	213465879	214365879	213465879	214365879
315264978	312475968	312475968	315264978	314256978	315284769
412375968	415263978	415263768	412395768	415273968	418293756
513284769	519273648	518273469	517283469	517293648	516273948
619273548	618293457	617283549	618243759	618243759	612345978
718293456	714283569	719243856	719253846	712384956	719253846
816243957	816253749	814293657	816293547	819263547	817263549
917253846	917243856	916253748	913274856	916283457	913245768
$\mathcal{F}121$	$\mathcal{F}122$	$\mathcal{F}123$	$\mathcal{F}124$	$\mathcal{F}125$	$\mathcal{F}126$

214365879	214365879	213465879	213465879	213465879	213465879
315284769	315264978	314256978	312475968	314256978	312475968
418293756	412395768	412395768	415263978	416273859	416293857
516273948	517283469	518293647	518243769	517293648	518273469
619243857	618293547	617283459	619273548	615283947	615243978
712354968	719253846	719243856	716253849	718263549	719283456
817263459	816243759	816273549	817293456	819243756	814253769
913254678	913274856	915263748	914283657	912345768	917263548
$\mathcal{F}127$	$\mathcal{F}128$	$\mathcal{F}129$	$\mathcal{F}130$	$\mathcal{F}131$	$\mathcal{F}132$

214365879	214365879	213465879	213465879	213465879	213465879
315284769	315264978	312495768	312495768	312475968	314256978
417293568	412395768	415283769	415283769	416253978	412395768
518273946	518273469	514293578	516243978	518273469	518263749
612345978	619253847	618273459	614273859	617293548	619273548
719243856	713294856	719243856	719263548	715263849	716243859
813264957	816243759	816293547	817293456	819243657	815293647
916253748	917283546	917253648	918253647	914283756	917283456
$\mathcal{F}133$	$\mathcal{F}134$	$\mathcal{F}135$	$\mathcal{F}136$	$\mathcal{F}137$	$\mathcal{F}138$

213465879	213465879	213465879	213465879	213465879	213465879
314256978	312475968	314256978	314256978	314256978	314256978
412395768	416253978	412395768	412395768	412375968	412395768
517263849	518243769	518293647	518273649	518273649	519263847
618243759	617283549	617283549	615293847	615283947	617293548
716293548	715293648	719243856	719283456	719263548	715283649
819273456	819273456	816273459	816243759	816293457	816273459
915283647	914263857	915263748	917263548	917243856	918243756
$\mathcal{F}139$	$\mathcal{F}140$	$\mathcal{F}141$	$\mathcal{F}142$	$\mathcal{F}143$	$\mathcal{F}144$
213465879	213465879	213465879	213465879	213465879	214365879
314256978	314256978	314265978	312495768	314256978	315274968
412395768	415293768	415273968	415283769	412395768	412385769
518273649	518263947	512374869	516243978	519283647	513294678
615293847	617243859	617283549	619273548	615293748	618243759
719263548	719283456	718293456	714263859	716243859	719263548
816243759	812364957	819243657	817293456	817263549	816253947
917283456	916273548	916253847	918253647	918273456	917283456
$\mathcal{F}145$	$\mathcal{F}146$	$\mathcal{F}147$	$\mathcal{F}148$	$\mathcal{F}149$	$\mathcal{F}150$
213465879	213465879	214365879	213465879	214365879	214365879
314256978	314256978	315264978	314256978	315284769	315264978
415273968	412395768	412395768	415273968	418293756	412375968
516293847	516293847	518243769	516293847	517263948	513284769
612374859	617283459	619253847	619283457	619243857	619273548
718263549	719263548	713294856	712364859	712354968	718253946
819243657	815273649	816273459	817263549	816273459	816293457
917283456	918243756	917283546	918243756	913254678	917243856
$\mathcal{F}151$	$\mathcal{F}152$	$\mathcal{F}153$	$\mathcal{F}154$	$\mathcal{F}155$	$\mathcal{F}156$
214365879	213465879	213465879	214365879	213465879	213465879
315264978	314256978	314265978	315264978	314265978	312495768
412385769	412395768	412395768	412385769	415273869	416253978
519243768	517283649	519273648	517243968	512374968	517243869
613284759	615293847	615293847	619283547	617293548	618273459
718293456	718263459	718243569	718293456	718243956	719263548
817253946	819243756	816253749	813274659	819253647	814293756
916273548	916273548	917283456	916253748	916283457	915283647
$\mathcal{F}157$	$\mathcal{F}158$	$\mathcal{F}159$	$\mathcal{F}160$	$\mathcal{F}161$	$\mathcal{F}162$
213465879	214365879	214365879	214365879	214365879	213465879
314256978	315264978	315294768	315264978	315264978	314256978
412375968	413295768	416273859	412395768	412375968	415273968
516273948	517243869	518243769	518243769	513284769	516293748
619243857	618273459	619283457	613284759	619243857	619283547
718293456	719283546	712394856	719253846	718253946	712384956
817263549	816253947	817263549	817293456	817293456	817243659
915283647	912374856	913254678	916273548	916273548	918263457
$\mathcal{F}163$	$\mathcal{F}164$	$\mathcal{F}165$	$\mathcal{F}166$	$\mathcal{F}167$	$\mathcal{F}168$

214365879	213465879	213465879	214365879	214365879	213465879
315264978	314256978	314256978	315264978	315264978	314256978
412385769	412375968	415273968	412375968	412375968	412395768
519243768	516283947	516293748	513274869	513284769	519283647
613284759	615273849	619283547	618243957	618293457	618273549
718253946	719263548	712384956	719283456	719253846	716243859
817293456	817293456	817263459	816293547	817243956	817293456
916273548	918243657	918243657	917253846	916273548	915263748
$\mathcal{F}169$	$\mathcal{F}170$	$\mathcal{F}171$	$\mathcal{F}172$	$\mathcal{F}173$	$\mathcal{F}174$
213465879	213465879	213465879	213465879	213465879	213465879
314256978	314256978	314256978	314256978	314256978	314256978
417293856	412375968	415273968	412395768	412395768	412395768
516283749	518273649	512374869	519263847	518263749	516293847
612394857	615283947	617283549	617283549	617293548	618243759
718243659	719263548	719243856	715293648	719243856	715283649
819263547	817293456	816293457	816273459	816273459	819273456
915273468	916243857	918253647	918243756	915283647	917263548
$\mathcal{F}175$	$\mathcal{F}176$	$\mathcal{F}177$	$\mathcal{F}178$	$\mathcal{F}179$	$\mathcal{F}180$
213465879	214365879	213465879	213465879	213465879	213465879
312475968	315294768	314256978	314256978	312475968	314256978
416253978	416273859	412375968	412395768	416253978	415273968
518273469	518243769	518263947	519273648	517283649	516293748
617283549	612394857	619283457	617283549	614293857	612384759
719243856	719283456	716293548	718293456	719263548	718263549
814293657	817263549	815273649	816243759	815273469	819243657
915263748	913254678	917243856	915263847	918243756	917283456
$\mathcal{F}181$	$\mathcal{F}182$	$\mathcal{F}183$	$\mathcal{F}184$	$\mathcal{F}185$	$\mathcal{F}186$
213465879	213465879	213465879	213465879	213465879	213465879
314256978	314256978	314275968	314265978	314256978	314256978
415273968	415293768	416293578	415293768	412395768	415293768
516273948	516273948	519263748	516273948	516293847	516273948
612384957	619283457	615283947	619283457	617283549	617283549
719283456	718263549	718253469	718253649	718243659	718243659
817243659	817243659	812364957	812354769	819273456	819263457
918263547	912384756	917243856	917243856	915263748	912384756
$\mathcal{F}187$	$\mathcal{F}188$	$\mathcal{F}189$	$\mathcal{F}190$	$\mathcal{F}191$	$\mathcal{F}192$
214365879	213465879	214365879	214365879	214365879	214365879
315264978	314256978	315264978	315294768	315274968	315264978
412395768	415273968	412395768	418253769	412385769	412375968
518243769	518293647	517283469	517263849	516293748	516273948
619253847	612374859	618293547	612394857	619283547	618293547
713284659	716283549	719243856	719283546	718263459	719253846
817293456	819263457	813274659	816273459	817253946	813245769
916273548	917243856	916253748	913245678	913245678	917283456
$\mathcal{F}193$	$\mathcal{F}194$	$\mathcal{F}195$	$\mathcal{F}196$	$\mathcal{F}197$	$\mathcal{F}198$

214365879	214365879	214365879	213465879	213465879	213465879
315274968	315284769	315264978	314265978	314265978	314265978
416253978	416253978	417253869	418293657	415293768	415293768
518293746	519243768	519283746	512384769	512384769	512384769
613284759	618293457	618243957	615283749	619283457	619273548
719263548	712384956	713294856	719253468	718253649	718243956
812345769	813274659	816273459	817243956	817243956	817253649
917243856	917263548	912354768	916273548	916273548	916283457
$\mathcal{F}$199	$\mathcal{F}$200	$\mathcal{F}$201	$\mathcal{F}$202	$\mathcal{F}$203	$\mathcal{F}$204
214365879	214365879	213465879	214365879	214365879	214365879
315264978	315294768	314256978	315264978	315264978	315264978
413285769	416253978	417283956	412385769	412385769	412375968
517243968	518243769	516293748	517293468	517243968	516273948
619253847	612384957	612384957	619283547	619253748	618293457
718293456	719263548	718243659	718243956	718293456	719243856
812374659	813274659	819263547	813274659	813274659	813254769
916273548	917283456	915273468	916253748	916283547	917283546
$\mathcal{F}$205	$\mathcal{F}$206	$\mathcal{F}$207	$\mathcal{F}$208	$\mathcal{F}$209	$\mathcal{F}$210
214365879	213465879	213465879	214365879	213465879	213465879
315264978	314256978	314256978	315274968	314256978	314256978
412385769	412395768	412395768	416253978	412395768	412375968
519243768	516293847	516293748	517283469	519283647	516273948
618273459	615283749	617243859	619243857	615273849	615293847
713294856	718243659	718263549	718293546	716293548	719283456
817253946	819273456	819273456	813264759	817263459	817263549
916283547	917263548	915283647	912374856	918243756	918243657
$\mathcal{F}$211	$\mathcal{F}$212	$\mathcal{F}$213	$\mathcal{F}$214	$\mathcal{F}$215	$\mathcal{F}$216
213465879	214365879	214365879	214365879	213465879	213465879
314256978	315264978	315284769	315294768	314256978	314256978
415273968	413295768	417293568	418253769	412375968	415273968
516293847	517283469	519263478	517263948	519263847	517283649
619283457	619273548	618243957	612384957	617293548	612384759
718243659	718243956	712384659	719283546	718243956	719263548
812374956	812374659	816253749	816273459	815273649	816293457
917263548	916253847	913274856	913245678	916283457	918243756
$\mathcal{F}$217	$\mathcal{F}$218	$\mathcal{F}$219	$\mathcal{F}$220	$\mathcal{F}$221	$\mathcal{F}$222
213465879	213465879	214365879	213465879	213465879	213465879
314256978	314256978	315274968	314256978	314256978	314256978
415293768	416273859	413285769	412375968	416293857	417293856
516273948	512374968	516243978	516293847	517243968	516273948
612384759	617293548	619253847	618243957	618273459	615283749
718263549	719283456	718293456	719263548	715283649	718243659
819243657	815263947	812374659	815273649	819263547	819263547
917283456	918243657	917263548	917283456	912374856	912345768
$\mathcal{F}$223	$\mathcal{F}$224	$\mathcal{F}$225	$\mathcal{F}$226	$\mathcal{F}$227	$\mathcal{F}$228

213465879	213465879	213465879	213465879	213465879	213465879
312495768	314256978	314256978	314256978	314256978	312495768
415283769	415293768	415273968	412395768	415273968	415283769
516243978	516273849	519263748	518263749	516293748	514263978
618293547	619283547	612384957	617283459	619283457	618273459
714263859	712394856	718243659	719243856	718243659	719253648
819273456	817243659	817293456	815293647	817263549	816293547
917253648	918263457	916283547	916273548	912384756	917243856
$\mathcal{F}229$	$\mathcal{F}230$	$\mathcal{F}231$	$\mathcal{F}232$	$\mathcal{F}233$	$\mathcal{F}234$

214365879	213465879	213465879	213465879	213465879	213465879
315284769	314265978	314275968	314265978	314256978	314265978
417293568	417283569	416253978	412395768	412395768	415293768
519263478	516273948	519263748	518273469	516283749	518273469
618253749	618293457	618293547	619253748	619273548	612394857
712384659	712384956	715283649	716283549	718243659	719243856
816243957	819253647	812345769	815293647	817293456	817253649
913274856	915243768	917243856	917243856	915263847	916283547
$\mathcal{F}235$	$\mathcal{F}236$	$\mathcal{F}237$	$\mathcal{F}238$	$\mathcal{F}239$	$\mathcal{F}240$

213465879	213465879	214365879	213465879	213465879	214365879
314265978	314256978	315284769	314256978	314256978	315264978
412375968	412375968	419263857	412375968	416283957	412395768
518293647	517263948	517243968	517293648	517293648	518293746
615273948	618273549	612345978	615283947	618273459	619283547
719243568	719283456	718293546	719243856	712384956	713254869
816253749	815293647	816253749	816273549	819263547	816273459
917283456	916243857	913274856	918263457	915243768	917243856
$\mathcal{F}241$	$\mathcal{F}242$	$\mathcal{F}243$	$\mathcal{F}244$	$\mathcal{F}245$	$\mathcal{F}246$

214365879	214365879	214365879	214365879	214365879	214365879
315284769	315264978	315264978	315264978	315274968	315264978
416253978	412395768	412385769	412385769	412385769	412395768
519243768	518273469	519243768	519243768	516293748	519273846
612384957	613284759	618253947	618253947	619283547	618293547
718293546	719243856	713294856	713284659	718263459	713254869
817263459	817293546	816273459	817293456	817243956	816243759
913274856	916253748	917283546	916273548	913254678	917283456
$\mathcal{F}247$	$\mathcal{F}248$	$\mathcal{F}249$	$\mathcal{F}250$	$\mathcal{F}251$	$\mathcal{F}252$

214365879	214365879	214365879	214365879	214365879	214365879
315284769	315264978	315284769	315264978	315274968	315264978
419263857	417253968	418263957	417253968	412385769	413285769
516293478	518293746	516293478	519273846	519263478	517293468
618243759	612384759	619253748	618243759	618293547	619273548
712354968	719283456	712354968	713294856	713284659	718243956
817253946	813245769	813274659	812354769	817243956	812374659
913274856	916273548	917243856	916283457	916253748	916253847
$\mathcal{F}253$	$\mathcal{F}254$	$\mathcal{F}255$	$\mathcal{F}256$	$\mathcal{F}257$	$\mathcal{F}258$

214365879	214365879	214365879	214365879	213465879	213465879
315284769	315274968	315284769	315264978	314256978	314256978
417293568	416253978	418293756	417253968	412395768	416273859
519273846	519283746	517263948	518293746	516293847	512374968
612345978	618293457	612345978	613284759	619273548	617293548
718243956	712354869	719253846	719243856	718243659	718243956
816253749	813264759	816273549	812345769	815263749	819263457
913264857	917243856	913245768	916273548	917283456	915283647
$\mathcal{F}$259	$\mathcal{F}$260	$\mathcal{F}$261	$\mathcal{F}$262	$\mathcal{F}$263	$\mathcal{F}$264
213465879	214365879	214365879	214365879	214365879	214365879
314256978	315264978	315264978	315264978	315284769	315264978
412375968	412375968	412385769	412395768	417293568	412395768
516273948	513284769	519273468	519283746	519273846	519273846
617283549	619273548	613284759	618273459	612345978	618243759
718293456	718243956	718243956	713254869	718243956	713254869
819243657	816293457	817293546	816293547	813264957	817293456
915263847	917253846	916253748	917243856	916253748	916283547
$\mathcal{F}$265	$\mathcal{F}$266	$\mathcal{F}$267	$\mathcal{F}$268	$\mathcal{F}$269	$\mathcal{F}$270
213465879	213465879	213465879	214365879	214365879	213465879
314256978	314256978	314256978	315264978	315264978	314275968
412395768	415273968	415273968	417253869	417283569	412395678
519283647	517293648	516293847	518273946	518273946	516283749
618273459	618243759	618243759	619283547	619253847	619243857
716293548	712384956	719283456	713294856	713294856	718253469
815263749	819263457	812364957	816243759	816243759	815293647
917243856	916283547	917263548	912345768	912345768	917263548
$\mathcal{F}$271	$\mathcal{F}$272	$\mathcal{F}$273	$\mathcal{F}$274	$\mathcal{F}$275	$\mathcal{F}$276
213465879	213465879	214365879	214365879	214365879	213465879
314265978	314256978	315264978	315284769	315274968	314265978
415293768	412395768	413295768	419253768	416253978	412395678
517283649	517293648	517243869	516293478	519283746	518273649
619273548	618243759	619283547	618243957	612384759	619283547
718243956	719283456	718253946	712384659	718293456	715243869
812345769	816273549	816273459	817263549	813245769	817293456
916253847	915263847	912374856	913274856	917263548	916253748
$\mathcal{F}$277	$\mathcal{F}$278	$\mathcal{F}$279	$\mathcal{F}$280	$\mathcal{F}$281	$\mathcal{F}$282
213465879	213465879	214365879	213465879	213465879	213465879
314275968	314256978	315264978	314256978	314256978	314256978
412385769	412395768	413295768	416273859	416273859	416273859
518293647	516273849	518273469	512374968	519243768	512374968
619253748	618243759	619253847	618293457	618293457	618243957
716283549	719283456	712394856	719263548	712394856	719263548
817243956	815293647	816243759	817243956	817263549	815293647
915263478	917263548	917283546	915283647	915283647	917283456
$\mathcal{F}$283	$\mathcal{F}$284	$\mathcal{F}$285	$\mathcal{F}$286	$\mathcal{F}$287	$\mathcal{F}$288

213465879	213465879	213465879	214365879	214365879	214365879
314256978	312495768	314295678	315264978	315264978	315284769
412375968	416273859	415273869	412375968	412395768	418293756
517293648	517293648	518263947	518273946	519283746	516273948
618273549	618253947	619283457	619283547	618293547	612354978
719283456	719283456	712354968	713254869	713254869	719253846
815263947	815243769	817243659	816293457	816273459	817263459
916243857	914263578	916253748	917243856	917243856	913245768
$\mathcal{F}289$	$\mathcal{F}290$	$\mathcal{F}291$	$\mathcal{F}292$	$\mathcal{F}293$	$\mathcal{F}294$
214365879	214365879	214365879	214365879	214365879	214365879
315264978	315294768	315264978	315284769	315274968	315264978
412385769	418253769	413285769	413295768	417253869	412395768
516293478	516273849	517243968	519273846	516293748	516293748
619253748	612394857	619253847	618243759	618243957	618273459
718243956	719283546	718293546	712354869	719283546	719243856
813264759	817263459	816273459	816253947	813264759	813254769
917283546	913245678	912374856	917283456	912345678	917283546
$\mathcal{F}295$	$\mathcal{F}296$	$\mathcal{F}297$	$\mathcal{F}298$	$\mathcal{F}299$	$\mathcal{F}300$
214365879	214365879	214365879	214365879	213465879	214365879
315264978	315264978	315264978	315284769	314256978	315264978
413285769	416273859	416273859	419263857	415293768	416283759
517293468	519283746	518293746	517293468	517283649	517243968
618253947	618253947	619283457	612354978	612384759	618293457
719243856	712354869	712354869	718253946	718243956	719253846
812374659	817293456	817243956	816243759	819263457	812354769
916273548	913245768	913254768	913274856	916273548	913274856
$\mathcal{F}301$	$\mathcal{F}302$	$\mathcal{F}303$	$\mathcal{F}304$	$\mathcal{F}305$	$\mathcal{F}306$
214365879	214365879	214365879	214365879	214365879	214365879
315274968	315274968	315264978	315264978	315284769	315264978
412385769	416293578	412395768	413295768	417293568	412375968
519283478	519283746	516293748	519283746	519263478	516273948
618253947	612394857	618273459	618253947	612384957	618293457
713294856	718263459	719283546	712354869	718243956	719283546
816243759	813254769	813254769	816273459	813274659	813254769
917283546	917243856	917243856	917243856	916253748	917243856
$\mathcal{F}307$	$\mathcal{F}308$	$\mathcal{F}309$	$\mathcal{F}310$	$\mathcal{F}311$	$\mathcal{F}312$
214365879	213465879	213465879	213465879	214365879	214365879
315294768	314256978	314275968	314256978	315264978	315264978
416253978	416283759	415283769	415293768	417253869	412385769
518243769	519263847	518293647	519263847	518273946	519273468
612384957	617293548	612354978	617283459	619283457	618253947
719283456	718243956	716253948	718243956	713294856	713294856
813274659	815273649	819263457	812364957	816243759	816243759
917263548	912345768	917243856	916273548	912354768	917283546
$\mathcal{F}313$	$\mathcal{F}314$	$\mathcal{F}315$	$\mathcal{F}316$	$\mathcal{F}317$	$\mathcal{F}318$

214365879	214365879	214365879	214365879	214365879	214365879
315284769	315264978	315264978	315284769	315264978	315274968
417293568	413285769	412385769	416293578	417283569	416253978
519263478	518293746	518273946	519243768	518293746	517283469
618243957	619253847	619283547	612384957	619243857	618293547
712384659	712345968	713245968	718253946	712394856	719243856
813274956	817243956	817293456	817263459	816273459	812374659
916253748	916273548	916253748	913274856	913254768	913264857
$\mathcal{F}$319	$\mathcal{F}$320	$\mathcal{F}$321	$\mathcal{F}$322	$\mathcal{F}$323	$\mathcal{F}$324
214365879	214365879	214365879	214365879	214365879	214365879
315284769	315274968	315284769	315264978	315274968	315274968
416253978	412385769	419263857	412375968	412385769	413285769
519243768	513294678	516293748	513274869	513294678	512394678
612384957	618243759	618273459	618253947	618253947	619253847
718263459	719283456	712354968	719283546	719283456	718293456
817293546	816253947	817253946	817293456	816243759	816243759
913274856	917263548	913245678	916243857	917263548	917263548
$\mathcal{F}$325	$\mathcal{F}$326	$\mathcal{F}$327	$\mathcal{F}$328	$\mathcal{F}$329	$\mathcal{F}$330
214365879	214365879	214365879	214365879	214365879	214365879
315264978	315284769	315264978	315264978	315274968	315264978
413285769	419263857	412385769	412385769	417253869	412385769
516293748	516243978	513294768	517293468	518263947	518293746
618273459	618253749	618243759	619283547	613294857	619273548
719243856	712345968	719283456	718253946	719283456	713245968
817253946	817293546	817253946	816243759	816243759	816253947
912354768	913274856	916273548	913274856	912354678	917283456
$\mathcal{F}$331	$\mathcal{F}$332	$\mathcal{F}$333	$\mathcal{F}$334	$\mathcal{F}$335	$\mathcal{F}$336
214365879	214365879	214365879	214365879	214365879	214365879
315264978	315264978	315264978	315284769	315284769	315274968
417253869	416273859	413285769	418263957	418293756	417253869
519283746	518293746	512394768	517293468	516273849	518263947
613294857	619283457	618273459	612354978	612345978	619283457
718243956	713254869	719243856	719243856	719263548	713294856
816273459	817243956	817293546	813274659	817253946	816243759
912354768	912354768	916253748	916253748	913245768	912354678
$\mathcal{F}$337	$\mathcal{F}$338	$\mathcal{F}$339	$\mathcal{F}$340	$\mathcal{F}$341	$\mathcal{F}$342
214365879	214365879	214365879	214365879	213465879	213465879
315284769	315264978	315284769	315284769	312495768	314275968
417253968	417253968	416253978	418293756	416253978	412385769
519263478	519283746	519243768	517243968	518293647	516243978
612384957	618273459	612384957	612345978	619273548	619283547
718293546	713294856	718293456	719253846	715243869	718253649
816243759	812354769	813274659	816273549	817263459	817293456
913274856	916243857	917263548	913264857	914283756	915263748
$\mathcal{F}$343	$\mathcal{F}$344	$\mathcal{F}$345	$\mathcal{F}$346	$\mathcal{F}$347	$\mathcal{F}$348

214365879	214365879	214365879	214365879	214365879	214365879
315264978	315274968	315264978	315284769	315274968	315264978
413285769	412385769	412375968	416273859	416283759	412385769
517243968	518293746	516273948	519243768	519263478	517243968
619273548	613245978	618293547	612394857	618243957	618293547
718293456	719283456	719283456	718263549	713294856	719283456
812374659	816253947	813245769	817293456	812354769	813274659
916253847	917263548	917253846	913254678	917253846	916253748
$\mathcal{F}349$	$\mathcal{F}350$	$\mathcal{F}351$	$\mathcal{F}352$	$\mathcal{F}353$	$\mathcal{F}354$

214365879	214365879	213465879	213465879	214365879	214365879
315264978	315284769	314275968	314256978	315284769	315264978
413295768	418273956	412385769	412395768	416253978	412385769
518273469	516293748	519243678	519263847	519243768	519283746
619283547	619243857	615283947	615283749	618293457	618253947
712394856	712345968	718263549	718243659	712384659	713245968
816243759	817263549	817293456	817293456	817263549	817293456
917253846	913254678	916253748	916273548	913274856	916273548
$\mathcal{F}355$	$\mathcal{F}356$	$\mathcal{F}357$	$\mathcal{F}358$	$\mathcal{F}359$	$\mathcal{F}360$

214365879	214365879	214365879	214365879	214365879	214365879
315264978	315284769	315264978	315264978	315264978	315264978
413295768	416253978	413295768	413295768	412385769	413285769
518273469	519243768	516273948	518273469	518273946	516293748
612384759	618293457	618243759	619283547	619253748	618273459
719283546	712384659	719253846	712384659	713245968	719253846
817243956	813274956	812354769	817243956	816293547	817243956
916253748	917263548	917283456	916253748	917283456	912354768
$\mathcal{F}361$	$\mathcal{F}362$	$\mathcal{F}363$	$\mathcal{F}364$	$\mathcal{F}365$	$\mathcal{F}366$

214365879	214365879	214365879	214365879	214365879	214365879
315264978	315274968	315264978	315264978	315274968	315264978
413295768	416283759	412385769	412385769	413285769	413285769
517243869	518263947	517293468	517293468	516243978	519243768
618253947	619243857	618253947	619253748	619253847	618253947
719283456	713254869	719283546	718243956	718263459	712384659
812374659	817293546	816243759	813274659	817293546	817293456
916273548	912345678	913274856	916283547	912374856	916273548
$\mathcal{F}367$	$\mathcal{F}368$	$\mathcal{F}369$	$\mathcal{F}370$	$\mathcal{F}371$	$\mathcal{F}372$

214365879	213465879	214365879	214365879	214365879	214365879
315284769	314275968	315264978	315274968	315264978	315284769
416273859	416293857	417253968	416253978	417253869	412395678
517243968	518243769	519273846	519263847	518273946	513274968
618293457	612354978	618243759	618243759	613294857	618243759
719263548	719283456	713294856	713294856	719283456	719253846
812374956	815263947	812345769	812345769	816243759	816293457
913254678	917253648	916283547	917283546	912354768	917263548
$\mathcal{F}373$	$\mathcal{F}374$	$\mathcal{F}375$	$\mathcal{F}376$	$\mathcal{F}377$	$\mathcal{F}378$

214365879	213465879	213465879	213465879	214365879	214365879
315274968	312475968	314275968	314275968	315264978	315264978
416253978	416253978	416253978	412395678	412385769	412385769
519283746	518243769	519263847	518293647	516273948	516293748
613294857	619273548	615293748	615283749	618243759	618273459
718263459	715283649	718243569	719263548	719283546	719283546
812354769	817293456	812364957	817253469	817293456	817243956
917243856	914263857	917283456	916243857	913254768	913254768
$\mathcal{F}$379	$\mathcal{F}$380	$\mathcal{F}$381	$\mathcal{F}$382	$\mathcal{F}$383	$\mathcal{F}$384

214365879	214365879	214365879	214365879	214365879	214365879
315264978	315264978	315274968	315284769	315264978	315274968
416283759	412385769	417263859	416253978	416283759	413285769
518273946	516273948	518243769	519243768	518273946	516293478
619243857	618243759	613294857	618273549	619253847	618253947
712354869	719283456	719283456	712384659	712354869	719263548
817293456	817293546	816253947	817293456	817293456	812374659
913254768	913254768	912354678	913264857	913245768	917243856
$\mathcal{F}$385	$\mathcal{F}$386	$\mathcal{F}$387	$\mathcal{F}$388	$\mathcal{F}$389	$\mathcal{F}$390

214365879	214365879	214365879	214365879	214365879	214365879
315274968	315264978	315264978	315274968	315274968	315274968
416253978	413285769	413285769	416283759	413285769	413285769
518293746	519273846	518273946	517263948	512394678	512394678
613284759	618293547	619253748	618293457	618293547	618243759
719243856	712345968	712345968	719253846	719243856	719263548
812345769	817243956	816293547	812354769	817263459	817293456
917263548	916253748	917243856	913245678	916253748	916253847
$\mathcal{F}$391	$\mathcal{F}$392	$\mathcal{F}$393	$\mathcal{F}$394	$\mathcal{F}$395	$\mathcal{F}$396

B

GENERATORS OF SIMPLE INDECOMPOSABLE FACTORIZATIONS

$2K_8$:
$\infty,0$	1,6	2,3	4,5	
$\infty,0$	1,5	2,4	3,6	

$4K_8$:
$\infty,0$	1,2	3,6	4,5	
$\infty,0$	1,4	2,3	5,6	
$\infty,0$	1,6	2,4	3,5	
$\infty,0$	1,5	2,4	3,6	

$2K_{10}$:
$\infty,0$	1,4	2,6	3,7	5,8
$\infty,0$	1,3	2,4	5,6	7,8

$4K_{10}$:
$\infty,0$	1,2	3,4	5,6	7,8
$\infty,0$	1,4	2,7	3,5	6,8
$\infty,0$	1,7	2,6	3,5	4,8
$\infty,0$	1,3	2,6	4,7	5,8

$5K_{10}$:
$\infty,0$	1,2	3,4	5,6	7,8
$\infty,0$	1,5	2,6	3,7	4,8
$\infty,0$	1,4	2,3	5,7	6,8
$\infty,0$	1,7	2,5	3,6	4,8
$\infty,0$	1,7	2,4	3,5	6,8

$3K_{12}$:

$\infty,0$	1,4	2,7	3,10	5,8	6,9
$\infty,0$	1,7	2,3	4,5	6,10	8,9
$\infty,0$	1,7	2,9	3,5	4,6	8,10

$4K_{12}$:

$\infty,0$	1,2	3,4	5,10	6,7	8,9
$\infty,0$	1,4	2,5	3,8	6,9	7,10
$\infty,0$	1,7	2,6	3,10	4,8	5,9
$\infty,0$	1,10	2,8	3,5	4,6	7,9

$6K_{12}$:

$\infty,0$	1,2	3,4	5,6	7,9	8,10
$\infty,0$	1,3	2,4	5,6	7,8	9,10
$\infty,0$	1,4	2,5	3,6	7,9	8,10
$\infty,0$	1,4	2,8	3,6	5,9	7,10
$\infty,0$	1,5	2,9	3,7	4,8	6,10
$\infty,0$	1,6	2,7	3,8	4,9	5,10

$8K_{12}$:

$\infty,0$	1,2	3,4	5,10	6,7	8,9
$\infty,0$	1,2	3,8	4,5	6,7	9,10
$\infty,0$	1,5	2,7	3,9	4,8	6,10
$\infty,0$	1,6	2,7	3,10	4,8	5,9
$\infty,0$	1,9	2,7	3,6	4,10	5,8
$\infty,0$	1,10	2,4	3,5	6,8	7,9
$\infty,0$	1,9	2,10	3,6	4,7	5,8
$\infty,0$	1,10	2,6	3,5	4,8	7,9

$9K_{12}$:

$\infty,0$	1,4	2,10	3,5	6,8	7,9
$\infty,0$	1,9	2,4	3,5	6,8	7,10
$\infty,0$	1,2	3,6	4,7	5,8	9,10
$\infty,0$	1,10	2,4	3,6	5,8	7,9
$\infty,0$	1,2	3,4	5,6	7,8	9,10
$\infty,0$	1,6	2,3	4,9	5,10	7,8
$\infty,0$	1,6	2,7	3,8	4,9	5,10
$\infty,0$	1,5	2,9	3,7	4,8	6,10
$\infty,0$	1,7	2,6	3,10	4,8	5,9

$3K_{14}$:

$\infty,0$	1,12	2,3	4,5	6,7	8,10	9,11
$\infty,0$	1,10	2,11	3,12	4,7	5,8	6,9
$\infty,0$	1,7	2,10	3,8	4,9	5,11	6,12

$8K_{14}$:

$\infty,0$	1,11	2,3	4,5	6,7	8,9	10,12
$\infty,0$	1,3	2,12	4,5	6,7	8,9	10,11
$\infty,0$	1,4	2,6	3,7	5,8	9,11	10,12
$\infty,0$	1,3	2,4	5,8	6,10	7,11	9,12
$\infty,0$	1,11	2,12	3,5	4,8	6,10	7,9
$\infty,0$	1,5	2,9	3,8	4,11	6,10	7,12
$\infty,0$	1,9	2,7	3,8	4,12	5,10	6,11
$\infty,0$	1,7	2,8	3,9	4,10	5,11	6,12

$9K_{14}$:

$\infty,0$	1,4	2,3	5,7	6,8	9,10	11,12
$\infty,0$	1,2	3,6	4,5	7,9	8,10	11,12
$\infty,0$	1,2	3,4	5,8	6,7	9,11	10,12
$\infty,0$	1,11	2,4	3,6	5,8	7,9	10,12
$\infty,0$	1,9	2,10	3,6	4,8	5,12	7,11
$\infty,0$	1,9	2,11	3,7	4,10	5,8	6,12
$\infty,0$	1,7	2,6	3,8	4,11	5,10	9,12
$\infty,0$	1,6	2,10	3,9	4,8	5,12	7,11
$\infty,0$	1,8	2,7	3,11	4,10	5,9	7,12

$10K_{14}$:

$\infty,0$	1,12	2,3	4,8	5,7	6,9	10,11
$\infty,0$	1,12	2,6	3,5	4,7	8,9	10,11
$\infty,0$	1,6	2,5	3,4	7,9	8,10	11,12
$\infty,0$	1,2	3,8	4,7	5,6	9,11	10,12
$\infty,0$	1,11	2,4	3,5	6,12	7,8	9,10
$\infty,0$	1,11	2,12	3,10	4,7	5,8	6,9
$\infty,0$	1,10	2,6	3,12	4,8	5,9	7,11
$\infty,0$	1,9	2,7	3,8	4,12	5,10	6,11
$\infty,0$	1,7	2,8	3,9	4,10	5,11	6,12
$\infty,0$	1,8	2,11	3,7	4,9	5,10	6,12

$5K_{16}$:

$\infty,0$	1,2	3,11	4,5	6,12	7,8	9,10	13,14
$\infty,0$	1,7	2,9	3,5	4,6	8,10	11,13	12,14
$\infty,0$	1,13	2,14	3,12	4,11	5,8	6,9	7,10
$\infty,0$	1,12	2,6	3,10	4,8	5,14	7,11	9,13
$\infty,0$	1,9	2,12	3,8	4,14	5,10	6,11	7,13

$7K_{16}$:

$\infty,0$	1,2	3,4	5,6	7,8	9,10	11,12	13,14
$\infty,0$	1,14	2,4	3,5	6,8	7,9	10,12	11,13
$\infty,0$	1,5	2,14	3,6	4,7	8,11	9,12	10,13
$\infty,0$	1,4	2,13	3,14	5,9	6,10	7,11	8,12
$\infty,0$	1,6	2,11	3,8	4,13	5,10	7,12	9,14
$\infty,0$	1,10,	2,8	3,12	4,9	5,14	6,11	7,13
$\infty,0$	1,8	2,9	3,10	4,11	5,12	6,13	7,14

$8K_{16}$:

$\infty,0$	1,2	3,4	5,6	7,8	9,10	11,12	13,14
$\infty,0$	1,3	2,4	5,7	6,8	9,11	10,12	13,14
$\infty,0$	1,13	2,14	3,5	4,7	6,9	8,11	10,12
$\infty,0$	1,4	2,6	3,7	5,9	8,12	10,13	11,14
$\infty,0$	1,5	2,12	3,13	4,8	6,10	7,11	9,14
$\infty,0$	1,11	2,7	3,13	4,9	5,10	6,12	8,14
$\infty,0$	1,10	2,8	3,9	4,13	5,11	6,12	7,14
$\infty,0$	1,8	2,9	3,10	4,11	5,12	6,13	7,14

$9K_{16}$:

$\infty,0$	1,2	3,4	5,6	7,8	9,10	11,12	13,14
$\infty,0$	1,14	2,4	3,5	6,8	7,9	10,11	12,13
$\infty,0$	1,3	2,5	4,7	6,9	8,10	11,13	12,14
$\infty,0$	1,4	2,14	3,6	5,8	7,11	10,13	9,12
$\infty,0$	1,12	2,13	2,14	4,8	5,9	6,10	7,11
$\infty,0$	1,6	2,11	3,8	4,14	5,10	7,12	9,13
$\infty,0$	1,6	2,7	3,12	4,10	5,11	8,13	9,14
$\infty,0$	1,7	2,11	3,9	4,10	5,12	6,13	8,14
$\infty,0$	1,8	2,9	3,10	4,11	5,12	6,13	7,14

$10K_{16}$:

$\infty,0$	1,2	3,4	5,6	7,8	9,10	11,12	13,14
$\infty,0$	1,3	2,4	5,6	7,8	9,11	10,12	13,14
$\infty,0$	1,3	2,4	5,7	6,9	8,10	11,13	12,14
$\infty,0$	1,13	2,5	3,6	4,7	8,11	9,12	10,14
$\infty,0$	1,4	2,5	3,6	7,11	8,12	9,13	10,14
$\infty,0$	1,5	2,12	3,7	4,8	6,11	9,13	10,14
$\infty,0$	1,11	2,7	3,8	4,13	5,10	6,12	9,14
$\infty,0$	1,7	2,11	3,9	4,14	5,10	6,12	8,13
$\infty,0$	1,8	2,9	3,11	4,10	5,14	6,12	7,13
$\infty,0$	1,8	2,9	3,10	4,11	5,12	6,13	7,14

$7K_{18}$:

$\infty,0$	1,2	3,12	4,5	6,7	8,9	10,11	13,14	15,16
$\infty,0$	1,16	2,11	3,5	4,6	7,9	8,10	12,14	13,15
$\infty,0$	1,4	2,10	3,7	5,8	6,9	11,14	12,15	13,16
$\infty,0$	1,15	2,6	3,16	4,13	5,9	7,11	8,12	10,14
$\infty,0$	1,13	2,14	3,8	4,9	5,11	6,12	7,16	10,15
$\infty,0$	1,7	2,14	3,8	4,13	5,11	6,12	9,15	10,16
$\infty,0$	1,11	2,9	3,10	4,13	5,12	6,16	7,14	8,15

$8K_{18}$:

$\infty,0$	1,2	3,4	5,6	7,8	9,10	11,12	13,14	15,16
$\infty,0$	1,3	2,4	5,7	6,8	9,11	10,12	13,15	14,16
$\infty,0$	1,15	2,16	3,6	4,7	5,8	9,12	10,13	11,14
$\infty,0$	1,5	2,6	3,7	4,8	9,13	10,14	11,15	12,16
$\infty,0$	1,6	2,14	3,15	4,9	5,10	7,12	11,16	8,13
$\infty,0$	1,7	2,13	3,9	4,15	5,11	6,12	8,14	10,16
$\infty,0$	1,8	2,12	3,10	4,11	5,15	6,13	7,14	9,16
$\infty,0$	1,9	2,10	3,11	4,12	5,13	6,14	7,15	8,16

$9K_{18}$:

$\infty,0$	1,2	3,4	5,6	7,8	9,10	11,12	13,14	15,16
$\infty,0$	1,16	2,4	3,5	6,8	7,9	10,12	11,13	14,15
$\infty,0$	1,15	2,16	3,6	4,7	5,8	9,11	10,13	12,14
$\infty,0$	1,5	2,15	3,7	4,8	6,10	9,12	11,14	13,16
$\infty,0$	1,14	2,6	3,15	4,8	5,10	7,12	9,13	11;16
$\infty,0$	1,12	2,7	3,14	4,16	5,10	6,11	8,13	9,15
$\infty,0$	1,12	2,8	3,14	4,10	5,11	6,16	7,13	9,15
$\infty,0$	1,11	2,9	3,10	4,14	5,12	6,13	7,16	8,15
$\infty,0$	1,9	2,10	3,11	4,12	5,13	6,14	7,15	8,16

$10K_{18}$:

∞,0	1,2	3,4	5,6	7,8	9,10	11,12	13,14	15,16
∞,0	1,2	3,4	5,7	6,8	9,11	10,12	13,15	14,16
∞,0	1,3	2,5	4,6	7,10	8,11	9,12	13,15	14,16
∞,0	1,5	2,16	3,6	4,7	8,12	9,13	10,14	11,15
∞,0	1,4	2,6	3,16	5,8	7,11	9,13	10,14	12,15
∞,0	1,9	2,14	3,15	4,16	5,10	6,11	7,12	8,13
∞,0	1,12	2,8	3,15	4,10	5,16	6,11	7,13	9,14
∞,0	1,12	2,9	3,10	4,15	5,11	6,16	7,13	8,14
∞,0	1,11	2,9	3,10	4,14	5,12	6,13	7,16	8,15
∞,0	1,9	2,10	3,11	4,12	5,13	6,14	7,15	8,16

$7K_{20}$:

∞,0	1,9	2,3	4,5	6,16	7,8	10,11	12,13	14,15	17,18
∞,0	1,3	2,10	4,6	5,7	8,17	9,11	12,14	13,15	16,18
∞,0	1,17	2,18	3,11	4,8	5,14	6,9	7,10	12,15	13,16
∞,0	1,5	2,6	3,11	4,8	7,17	9,13	10,14	12,16	15,18
∞,0	1,15	2,8	3,9	4,14	5,17	6,17	7,12	11,16	13,18
∞,0	1,14	2,11	3,16	4,17	5,10	6,12	7,18	8,13	9,15
∞,0	1,8	2,12	3,10	4,15	5,17	6,13	7,14	9,16	11,18

$8K_{20}$:

∞,0	1,10	2,3	4,5	6,7	8,9	11,12	13,14	15,16	17,18
∞,0	1,3	2,4	5,14	6,8	7,9	10,12	11,13	15,17	16,18
∞,0	1,4	2,18	3,6	5,15	7,10	8,11	9,12	13,16	14,17
∞,0	1,16	2,17	3,7	4,13	5,9	6,10	8,12	11,15	14,18
∞,0	1,6	2,7	3,8	4,9	5,14	10,15	11,16	12,17	13,18
∞,0	1,14	2,8	3,9	4,17	5,11	6,15	7,13	10,16	12,18
∞,0	1,13	2,14	3,10	4,11	5,12	6,18	7,17	8,15	9,16
∞,0	1,9	2,13	3,11	4,12	5,16	6,14	7,15	8,17	10,18

$9K_{20}$:

∞,0	1,2	3,4	5,6	7,8	9,10	11,12	13,14	15,16	17,18
∞,0	1,18	2,4	3,5	6,8	7,9	10,12	11,13	14,16	15,17
∞,0	1,4	2,5	3,6	7,10	8,11	9,12	13,16	14,17	15,18
∞,0	1,5	2,17	3,7	4,8	6,10	9,13	11,15	12,16	14,18
∞,0	1,15	2,16	3,17	4,18	5,10	6,11	7,12	8,13	9,14
∞,0	1,14	2,8	3,16	4,10	5,18	6,12	7,13	9,15	11,17
∞,0	1,8	2,9	3,15	4,16	5,12	6,13	7,14	10,17	11,18
∞,0	1,12	2,10	3,11	4,15	5,13	6,14	7,18	8,16	9,17
∞,0	1,10	2,11	3,12	4,13	5,14	6,15	7,16	8,17	9,18

$10K_{20}$:

∞,0	1,2	3,4	5,6	7,8	9,10	11,12	13,14	15,16	17,18
∞,0	1,3	2,4	5,7	6,8	9,10	11,13	12,14	15,17	16,18
∞,0	1,17	2,18	3,6	4,7	5,8	9,11	10,13	12,15	14,16
∞,0	1,5	2,17	3,18	4,7	6,9	8,12	10,14	11,15	13,16
∞,0	1,5	2,17	3,7	4,18	6,11	8,13	9,14	10,15	12,16
∞,0	1,6	2,16	3,17	4,9	5,11	7,13	8,14	10,15	12,18
∞,0	1,7	2,14	3,9	4,17	5,11	6,13	8,15	10,16	12,18
∞,0	1,8	2,9	3,14	4,16	5,12	6,13	7,15	10,17	11,18
∞,0	1,9	2,13	3,11	4,12	5,14	6,17	7,15	8,16	10,18
∞,0	1,10	2,11	3,12	4,13	5,14	6,15	7,16	8,17	9,18

$7K_{22}$:

∞,0	1,2	3,4	5,6	7,16	8,18	9,17	10,11	12,13	14,15	19,20
∞,0	1,13	2,12	4,5	4,6	7,15	8,10	9,11	14,16	17,19	18,20
∞,0	1,4	2,12	3,6	5,9	7,20	8,11	10,19	13,16	14,17	15,18
∞,0	1,14	2,19	3,20	4,8	5,9	6,18	7,17	10,13	11,15	12,16
∞,0	1,6	2,14	3,9	4,20	5,10	7,15	8,18	11,16	12,17	13,19
∞,0	1,17	2,10	3,19	4,16	5,11	6,12	7,13	8,18	9,15	14,20
∞,0	1,15	2,10	3,17	4,18	5,12	6,13	7,14	8,19	9,16	11,20

$10K_{22}$:

∞,0	1,2	3,4	5,6	7,8	9,10	11,12	13,14	15,16	17,18	19,20
∞,0	1,3	2,4	5,7	6,8	9,11	10,12	13,15	14,16	17,19	18,20
∞,0	1,4	2,5	3,7	6,9	8,12	10,13	11,14	15,18	16,19	17,20
∞,0	1,5	2,19	3,20	4,8	6,10	7,11	9,13	12,16	14,17	15,18
∞,0	1,17	2,7	3,8	4,9	5,20	6,12	10,15	11,16	13,18	14,19
∞,0	1,16	2,18	3,19	4,10	5,20	6,12	7,13	8,14	9,15	11,17
∞,0	1,9	2,16	3,17	4,11	5,19	6,13	7,14	8,15	10,18	12,20
∞,0	1,9	2,15	3,11	4,17	5,12	6,19	7,14	8,16	10,18	13,20
∞,0	1,10	2,13	3,12	4,16	5,14	6,15	7,19	8,17	9,18	11,20
∞,0	1,11	2,12	3,13	4,14	5,15	6,17	7,16	8,18	9,19	10,20

$7K_{24}$:

∞,0	1,2	3,4	5,16	6,15	7,8	9,17	10,20	11,12	13,14	18,19	21,22
∞,0	1,3	2,4	5,15	6,8	7,16	9,21	10,12	11,13	14,22	17,19	18,20
∞,0	1,14	2,11	3,7	4,16	5,8	6,21	9,12	10,13	15,18	17,20	19,22
∞,0	1,5	2,21	3,12	4,15	6,19	7,11	8,16	9,13	10,14	17,20	18,22
∞,0	1,6	2,20	3,15	4,10	5,18	7,21	8,13	9,14	11,19	12,17	16,22
∞,0	1,13	2,7	3,9	4,21	5,22	6,16	8,17	10,15	11,19	12,18	14,20
∞,0	1,16	2,9	3,10	4,13	5,17	6,22	7,20	8,15	11,18	12,19	14,21

$10K_{24}$:

∞,0	1,2	3,4	5,6	7,18	8,9	10,11	12,13	14,15	16,17	19,20	21,22
∞,0	1,3	2,4	5,7	6,8	9,11	10,21	12,14	13,15	16,18	17,19	20,22
∞,0	1,4	2,5	3,14	6,9	7,10	8,11	12,15	13,16	17,20	18,21	19,22
∞,0	1,5	2,21	3,22	4,8	6,10	7,11	9,20	12,16	13,17	14,18	15,19
∞,0	1,19	2,14	3,8	4,9	5,10	6,11	7,12	13,18	15,20	16,21	17,22
∞,0	1,7	2,8	3,9	4,10	5,11	6,17	12,18	13,19	14,20	15,21	16,22
∞,0	1,8	2,18	3,10	4,11	5,20	6,17	7,14	9,16	12,19	13,21	15,22
∞,0	1,9	2,10	3,18	4,11	5,20	6,17	7,15	8,16	12,19	13,21	14,22
∞,0	1,15	2,11	3,12	4,16	5,14	6,20	7,21	8,17	9,18	10,19	13,22
∞,0	1,14	2,12	3,13	4,17	5,15	6,16	7,19	8,18	9,22	10,20	11,21

$7K_{26}$:

∞,0	1,16	2,3	4,18	5,6	7,8	9,17	10,19	11,24
					12,13	14,15	20,21	22,23
∞,0	1,3	2,4	5,17	6,16	7,9	8,10	11,13	12,23
					14,22	15,24	18,20	19,21
∞,0	1,22	2,24	3,13	4,7	5,8	6,20	9,17	10,23
					11,14	12,21	15,18	16,19
∞,0	1,13	2,6	3,17	4,14	5,9	7,10	8,16	11,15
					12,21	18,22	19,23	20,24
∞,0	1,9	2,7	3,16	4,23	5,24	6,11	8,18	10,21
					12,17	13,22	14,19	15,20
∞,0	1,7	2,8	3,17	4,10	5,11	6,19	9,15	12,20
					13,22	14,24	16,21	18,23
∞,0	1,9	2,16	3,21	4,17	5,23	6,24	7,14	8,15
					10,19	11,18	12,22	13,20

$7K_{28}$:

∞,0	1,2	3,4	5,21	6,16	7,8	9,18	10,22	11,12
				13,26	14,15	17,25	19,20	23,24
∞,0	1,26	2,14	3,21	4,23	5,19	6,8	7,24	9,25
				10,12	11,13	15,17	16,18	20,22
∞,0	1,4	2,10	3,6	5,19	7,17	8,26	9,21	11,14
				12,16	13,24	15,18	20,23	22,25
∞,0	1,5	2,14	3,7	4,8	6,15	9,13	10,18	11,25
				12,23	16,26	17,21	19,22	20,24
∞,0	1,6	2,14	3,8	4,23	5,16	7,12	9,22	10,20
				11,17	13,18	15,24	19,25	21,26
∞,0	1,22	2,17	3,19	4,25	5,24	6,12	7,21	8,14
				9,18	10,15	11,16	13,23	20,26
∞,0	1,14	2,10	3,19	4,11	5,12	6,26	7,25	8,18
				9,21	13,20	15,22	16,23	17,24

GENERATORS OF NONSIMPLE INDECOMPOSABLE FACTORIZATIONS

$5K_8$:

$\infty,0$	1,2	3,4	5,6				
$\infty,0$	1,3	2,5	4,6				
$\infty,0$	1,5	2,6	3,4 twice				
$\infty,0$	1,6	2,4	3,5				

$6K_8$:

$\infty,0$	1,2	3,4	5,6 twice				
$\infty,0$	1,4	2,5	3,6				
$\infty,0$	1,3	2,5	4,6 three times				

$12K_16$:

$\infty,0$	1,14	2,3	4,5	6,7	8,9	10,11	12,13 twice
$\infty,0$	1,4	2,7	3,14	5,13	6,12	8,10	9,11 five times
$\infty,0$	1,6	2,14	3,7	4,10	5,13	8,11	9,12
$\infty,0$	1,10	2,5	3,14	4,9	6,13	7,11	8,12
$\infty,0$	1,7	2,12	3,14	4,9	5,10	6,13	8,11
$\infty,0$	1,11	2,8	3,14	4,10	5,12	6,9	7,13
$\infty,0$	1,9	2,10	3,7	4,14	5,13	6,12	8,11

REFERENCES

[1] N. Alon and K. A. Berman, Regular hypergraphs, Gordon's lemma, Steinitz' lemma and invariant theory. *Journal of Combinatorial Theory* 43A (1986), 91–97.

[2] B. Alspach, A 1-factorization of the line graphs of complete graphs. *Journal of Graph Theory* 6 (1982), 411–415.

[3] B. Alspach and J. C. George, One-factorizations of tensor products of graphs. In *Topics in Combinatorics and Graph Theory* (Physica-Verlag, 1990), 41–46.

[4] B. Alspach, K. Heinrich and G. Liu, Orthogonal factorizations of graphs. In *Contemporary Design Theory* (Wiley, New York, 1992), 13–40.

[5] B. Alspach, P. Schellenberg, D. Stinson and D. Wagner, The Oberwolfach problem and factors of uniform odd length cycles. *Journal of Combinatorial Theory* 52A (1989), 20–43.

[6] B. A. Anderson, Finite topologies and Hamiltonian paths. *Journal of Combinatorial Theory* 14B (1973), 87–93.

[7] B. A. Anderson, A perfectly arranged Room square. *Congressus Numerantium* 8 (1973), 141–150.

[8] B. A. Anderson, A class of starter induced one-factorizations. *Lecture Notes in Mathematics* 406 (1974), 180–185.

[9] B. A. Anderson, Some perfect one-factorizations. *Congressus Numerantium* 17 (1976), 79–91.

[10] B. A. Anderson, Symmetry groups of some perfect one-factorizations of complete graphs. *Discrete Mathematics* 18 (1977), 227–234.

[11] B. A. Anderson, M. M. Barge and D. Morse, A recursive construction of asymmetric 1-factorizations. *Aequationes Mathematicae* 15 (1977), 201–211.

225

[12] B. A. Anderson and D. Morse, Some observations on starters. *Congressus Numerantium* 10 (1974), 229–235.

[13] D. Archdeacon and J. H. Dinitz, Constructing indecomposable 1-factorizations of the complete multigraph. *Discrete Mathematics* 92 (1991), 9–19.

[14] A. H. Baartmans and W. D. Wallis, Indecomposable factorizations of multigraphs. *Discrete Mathematics* 78 (1989), 37–43.

[15] Zs. Baranyai, On the factorization of the complete uniform hypergraph. In *Infinite and Finite Sets, Proceedings of the Conference, Keszthely, 1973 (Colloquia Mathematica Societas János Bolyai* 10, Volume I, North-Holland, 1975), 91–108.

[16] I. R. Beaman and W. D. Wallis, A skew Room square of side 9. *Utilitas Mathematica* 8 (1975), 382.

[17] A. F. Beecham and A. C. Hurley, A scheduling problem with a simple graphical solution. *Journal of the Australian Mathematical Society* 21B (1980), 486–495.

[18] L. W. Beineke and R. J. Wilson, On the edge-chromatic number of a graph. *Discrete Mathematics* 5 (1973), 15–20.

[19] M.B. Beintema, J.T. Bonn, R.W. Fitzgerald and J.L. Yucas, Orderings of finite fields and balanced tournaments. *Ars Combinatoria* (to appear).

[20] C. Berge, *Graphs and Hypergraphs*. (North-Holland, 1973.)

[21] N.L. Biggs, E.K. Lloyd and R.J. Wilson, *Graph Theory, 1736-1936*. (Clarendon Press, Oxford, 1976.)

[22] D. C. Blest and D. G. Fitzgerald, Scheduling sports competitions with a given distribution of times. *Discrete Applied Mathematics* 22 (1988-89), 9–19.

[23] J.A. Bondy and V. Chvátal, A method in graph theory. *Discrete Mathematics* 15 (1976), 111–136.

[24] J. T. Bonn, *Combinatorial Objects from Ordering the Elements of a Finite Field*. (Ph. D. Dissertation, Southern Illinois University, 1996.)

[25] J. Bosák, *Decompositions of Graphs*. (Kluwer, Dordrecht, 1990.)

[26] R. C. Bose, On the construction of balanced incomplete block designs. *Annals of Eugenics* 9 (1939), 353–399.

[27] R. C. Bose, S. S. Shrikhande and E. T. Parker, Further results on the construction of mutually orthogonal Latin squares and the falsity of Euler's conjecture. *Canadian Journal of Mathematics* 12 (1960), 189–203.

[28] F. C. Bussemaker, S. Čobelijić, D. M. Cvetković and J. J. Seidel, Computer investigation of cubic graphs. *Report 76-WSK-01, Technological University Eindhoven* (1976).

[29] L. Caccetta and S. Mardiyono, On maximal sets of one-factors. *Australasian Journal of Combinatorics* 1 (1990), 5–14.

[30] L. Caccetta and S. Mardiyono, Premature sets of one-factors. *Australasian Journal of Combinatorics* 5 (1992), 229–252.

[31] L. Caccetta and S. Mardiyono, On the existence of almost-regular graphs without one-factors. *Australasian Journal of Combinatorics* 9 (1994), 243–260.

[32] L. Caccetta and S. Mardiyono, On the existence of almost-regular graphs with exactly one one-factor. *Journal of Combinatorial Mathematics and Combinatorial Computing* 21 (1996), 161–177.

[33] L. Caccetta and S. Mardiyono, On graphs without one-factors. *Australasian Journal of Combinatorics* (to appear).

[34] L. Caccetta and W. D. Wallis, Maximal sets of deficiency three. *Congressus Numerantium* 23 (1979), 217–227.

[35] P. Cameron, Minimal edge-colouring of complete graphs. *Journal of the London Mathematical Society* (2) 11 (1975), 337–346.

[36] P. Cameron, *Parallelisms in Complete Designs.* (Cambridge University Press, 1976.)

[37] A. G. Chetwynd and A. J. W. Hilton, Regular graphs of high degree are 1-factorizable. *Proceedings of the London Mathematical Society* (3) 50 (1985), 193–206.

[38] A. G. Chetwynd and A. J. W. Hilton, The edge-chromatic class of regular graphs of degree 4 and their complements. *Discrete Applied Mathematics* 16 (1987), 125–134.

[39] A. G. Chetwynd and A. J. W. Hilton, 1-factorizing regular graphs of high degree – an improved bound. *Discrete Mathematics* 75 (1989), 103–112.

[40] C. J. Colbourn, M. J. Colbourn and A. Rosa, Indecomposable one-factorizations of the complete multigraph. *Journal of the Australian Mathematical Society*, 39A (1985), 334–343.

[41] C. J. Colbourn and G. Nonay, A golf design of order 11. *Journal of Statistical Planning and Inference* (to appear).

[42] E. Cousins and W. D. Wallis, Maximal sets of one-factors. In *Combinatorial Mathematics III* (*Lecture Notes in Mathematics* 452, Springer-Verlag, 1975), 90–94.

[43] A. Cruse, On embedding incomplete symmetric Latin squares. *Journal of Combinatorial Theory* 16A (1974), 18–22.

[44] A. B. Cruse, A note on one-factors in certain regular multigraphs. *Discrete Mathematics* 18 (1977), 213–216.

[45] E. E. Davis, Pairing teams. *Mathematics Magazine* 32 (1958-59), 99–100.

[46] L. E. Dickson and F. H. Safford, Solution to problem 8 (group theory). *American Mathematical Monthly* 13 (1906), 150–151.

[47] J. H. Dinitz, Room n-cubes of low order. *Journal of the Australian Mathematical Society* 36A (1984), 237–252.

[48] J. H. Dinitz, The existence of Room 5-cubes. *Journal of the Combinatorial Theory* 45A (1987), 125–138.

[49] J. H. Dinitz, Some perfect Room squares. *Journal of Combinatorial Mathematics and Combinatorial Computing* 2 (1987), 29–36.

[50] J. H. Dinitz and D. K. Garnick, There are 23 non-isomorphic perfect one-factorizations of K_{14}. *Journal of Combinatorial Designs* 4 (1996), 1–4.

[51] J. H. Dinitz, D. K. Garnick and B. D. McKay, There are 526,915,620 non-isomorphic one-factorizations of K_{12}. *Journal of Combinatorial Designs* 2 (1994), 273–285.

[52] J. H. Dinitz and D. R. Stinson, The construction and uses of frames. *Ars Combinatoria* 10 (1980), 34–54.

[53] J. H. Dinitz and D. R. Stinson, A fast algorithm for finding strong starters. *SIAM Journal of Algebraic and Discrete Methods* 2 (1981), 50–56.

[54] J. H. Dinitz and D. R. Stinson, On nonisomorphic Room squares. *Proceedings of the American Mathematical Society* 89 (1983), 175–181.

[55] J. H. Dinitz and D. R. Stinson, A hill–climbing algorithm for the construction of one-factorizations and Room squares. *SIAM Journal of Algebraic and Discrete Methods* 8 (1987), 430–438.

[56] J. H. Dinitz and D. R. Stinson, Some new perfect one-factorizations from starters in finite fields. *Journal of Graph Theory* 13 (1989), 405–415.

[57] J. H. Dinitz and D. R. Stinson, Room squares and related designs. In *Contemporary Design Theory* (Wiley, New York, 1992), 137–204.

[58] J. H. Dinitz and W. D. Wallis, Four orthogonal one-factorizations on ten points. *Annals of Discrete Mathematics* 26 (1985), 143–150.

[59] J. H. Dinitz and W. D. Wallis, Trains: an invariant for one-factorizations. *Ars Combinatoria* 32 (1991), 161–180.

[60] G. A. Dirac, Some theorems on abstract graphs. *Proceedings of the London Mathematical Society* (Series 3) 2 (1952), 69–81.

[61] J. Doyen and R. M Wilson, Embeddings of Steiner triple systems. *Discrete Mathematics* 5 (1973), 229–239.

[62] P. Erdös and I. Kaplansky, The asymptotic number of Latin rectangles. *American Journal of Mathematics* 68 (1946), 230–236.

[63] L. Euler, Solutio Problematis ad geometriam situs pertinentis. *Commentarii Academiae Scientiarum Imperialis, Petropolitanae* 8 (1736), 128–140.

[64] G. Fan, Covering weighted graphs by even subgraphs. *Journal of Combinatorial Theory* 49B (1990), 137–141.

[65] N. J. Finizio, Tournament designs balanced with respect to several bias categories. *Bulletin of the Institute of Combinatorics and its Applications* 9 (1993), 69–95.

[66] S. Fiorini and R. J. Wilson, *Edge-Colourings of Graphs*. (Pitman, London, 1977.)

[67] J. Folkman and J.R. Fulkerson, Edge colorings in bipartite graphs. In *Combinatorial Mathematics and its Applications* (University of North Carolina Press, 1969), 561–577.

[68] J.-C. Fournier, Colorations des arêtes d'un graphe. *Cahiers du CERO* 15 (1973), 311–314.

[69] J. E. Freund, Round robin mathematics. *American Mathematical Monthly* 63 (1956), 112–114.

[70] C. M. Fu and H. L. Fu, The mutual intersections of three distinct 1-factorizations. *Ars Combinatoria* 28 (1989), 55–64.

[71] C. M. Fu and H. L. Fu, On the intersection of Latin squares with holes. *Ars Combinatoria* 28 (1989), 55–64.

[72] H. L. Fu, On the constructions and applications of two 1-factorizations with prescribed intersections. *Tamking Journal of Mathematics* 16 (1985), 117–124.

[73] E. N. Gelling, *On one-factorizations of a complete graph and the relationship to round-robin schedules.* (MA Thesis, University of Victoria, Canada, 1973.)

[74] E. N. Gelling and R. E. Odeh, On 1-factorizations of the complete graph and the relationship to round-robin schedules. *Congressus Numerantium* 9 (1974), 213–221.

[75] J. C. George, *1-Factorizations of Tensor Products of Graphs.* (PhD Dissertation, University of Illonois, USA, 1991.)

[76] J. E. Graver, A survey of the maximum depth problem for indecomposable exact covers. In *Infinite and Finite Sets, Proceedings of the Conference, Keszthely, 1973 (Colloquia Mathematica Societas János Bolyai*, North-Holland, Amsterdam, 1973), 731–743.

[77] T. S. Griggs and A. Rosa, An invariant for one-factorizations of the complete graph. *Ars Combinatoria* 42 (1996), 77–88.

[78] K. B. Gross, Equivalence of Room designs II. *Journal of Combinatorial Theory* 17A (1974), 299–316.

[79] K. B. Gross, A multiplication theorem for strong starters. *Aequationes Mathematicae* 11 (1974), 169–173.

[80] R. P. Gupta, The chromatic index and the degree of a graph. *Notices of the American Mathematical Society* 13 (1966), 719 (Abstract 66T-429).

[81] P. Hall, On representatives of subsets. *Journal of the London Mathematical Society* 10 (1935), 26–30.

[82] W.R. Hamilton, *The Icosian Game* (leaflet, Jaques and Son, 1859). Reprinted in [21], 33–35.

[83] G. H. Hardy and E. M. Wright. *The Theory of Numbers.* (Oxford University Press, New York, 1938.)

[84] A. Hartman, Tripling quadruple systems. *Ars Combinatoria* 10 (1980), 255–309.

[85] A. Hartman and A. Rosa, Cyclic one-factorizations of the complete graph. *European Journal of Combinatorics* 6 (1985), 45–48.

[86] A. J. W. Hilton, Factorizations of regular graphs of high degree. *Journal of Graph Theory* 9 (1985), 193–196.

[87] P. Himelwright, W. D. Wallis and J. E. Williamson, On one-factorizations of compositions of graphs. *Journal of Graph Theory* 6 (1982), 75–80; Erratum, *ibid.* 8 (1984), 185–186.

[88] P. E. Himelwright and J. E. Williamson, On 1-factorability and edge-colorability of cartesian products of graphs. *Elemente der Mathematik* 29 (1974), 66–67.

[89] J. D. Horton, Quintuplication of Room squares. *Aequationes Mathematicae* 7 (1971), 243–245.

[90] J. D. Horton, Room designs and one-factorizations. *Aequationes Mathematicae* 22 (1983), 56–63.

[91] J. D. Horton and W. D. Wallis, Factoring the cartesian product of a cubic graph and a triangle. (to appear).

[92] F. K. Hwang, How to design round robin schedules. In *Combinatorics, Computing and Complexity (Tianjing and Beijing, 1988)* (Kluwer, Dordrecht, 1989), 142–160.

[93] E. C. Ihrig, Symmetry groups related to the construction of perfect one-factorizations of K_{2n}. *Journal of Combinatorial Theory* 40B (1986), 121–151.

[94] E. C. Ihrig, Symmetry groups of perfect one factorizations of K_{2n}. *Congressus Numerantium* 60 (1987), 105–130.

[95] E. C. Ihrig, The structure of the symmetry groups of perfect 1-factorizations of K_{2n}. *Journal of Combinatorial Theory* 47B (1989), 307–329.

[96] E. C. Ihrig, E. Seah and D. R. Stinson, A perfect one-factorization of K_{50}. *Journal of Combinatorial Mathematics and Combinatorial Computing* 1 (1987), 217–219.

[97] Q. Kang, *Large Sets of Triple Systems and Related Designs.* (Science Press, New York, 1995).

[98] P. Katerinis, Maximum matchings in a regular graph of specified connectivity and bounded order. *Journal of Graph Theory* 11 (1987), 53–58.

[99] T. P. Kirkman, On a problem in combinations. *Combridge and Dublin Mathematics Journal* 2 (1847), 191–204.

[100] T. P. Kirkman, On the representation of polyedra. *Philosophical Transactions of the Royal Society, London* 146 (1856), 413–418.

[101] M. Kobayashi, On perfect one-factorization of the complete graph K_{2p}. *Graphs and Combinatorics* 5 (1989), 351–353.

[102] M. Kobayashi, H. Awoki, Y. Nakazaki and G. Nakamura, A perfect one-factorization of K_{36}. *Graphs and Combinatorics* 5 (1989), 243–244.

[103] M. Kobayashi and Kiyasu-Zen'iti, Perfect one-factorizations of K_{1332} and K_{6860}. *Journal of Combinatorial Theory* 51A (1989), 314–315.

[104] G. Korchmáros, Cyclic one-factorization with an invariant one-factor of the complete graph. *Ars Combinatoria* 27 (1989), 133–138.

[105] A. Kotzíg, Problems and recent results on 1-factorizations of cartesian products of graphs. *Congressus Numerantium* 21 (1978), 457–460.

[106] A. Kotzíg, 1-factorizations of cartesian products of regular graphs. *Journal of Graph Theory* 3 (1979), 23–34.

[107] C. C. Lindner, E. Mendelsohn and A. Rosa, On the number of 1-factorizations of the complete graph. *Journal of Combinatorial Theory* 20A (1976), 265–282.

[108] C. C. Lindner and W. D. Wallis, A note on one-factorizations having a prescribed number of edges in common. *Annals of Discrete Mathematics* 12 (1982), 203–209.

[109] L. Lovász, Three short proofs in graph theory. *Journal of Combinatorial Theory* 19B (1975), 269–271.

[110] L. Lovász and M. D. Plummer, *Matching Theory.* (North-Holland, Amsterdam, 1986.)

[111] E. S. Mahmoodian, On edge-colorability of cartesian products of graphs. *Canadian Mathematical Bulletin* 24 (1981), 107–108.

[112] H. B. Mann and H. J. Ryser, Systems of distinct representatives. *American Mathematical Monthly* 60 (1953), 397–401.

[113] S. Mardiyono, *Factors in Regular and Almost-Regular Graphs*. (PhD Thesis, Curtin University of Technology, Australia, 1995.)

[114] E. Mendelsohn and A. Rosa, On some properties of 1-factorizations of complete graphs. *Congressus Numerantium* 24 (1979), 739–752.

[115] E. Mendelsohn and A. Rosa, One-factorizations of the complete graph – A survey. *Journal of Graph Theory* 9 (1985), 43–65.

[116] B. Mohar, On edge-colorability of products of graphs. *Publications de l'Institut Mathématique* (Nouvelle Série) 36 (50) (1984), 13–16.

[117] B. Mohar and T. Pisanski, Edge-coloring of a family of regular graphs. *Publications de l'Institut Mathématique* (Nouvelle Série) 33 (47) (1983), 157–162.

[118] B. Mohar, T. Pisanski and J. Shawe-Taylor, Edge-coloring of composite regular graphs. *Colloquia Mathematica Societas János Bolyai* 37 (1981), 591–600.

[119] R. C. Mullin and E. Nemeth, An existence theorem for Room squares. *Canadian Mathematical Bulletin* 12 (1969), 493–497.

[120] G. Nakamura, Dudeney's round table problem and the edge-coloring of the complete graph. (In Japanese.) *Sūgaku Seminar* 15 (1975), 24–29.

[121] C. St.J. A. Nash-Williams, Edge-disjoint spanning trees of finite graphs. *Journal of the London Mathematical Society* 36 (1961), 445–450.

[122] O. Ore, Note on Hamilton circuits. *American Mathematical Monthly* 67 (1960), 55.

[123] O. Ore, Hamilton connected graphs. *Journal de Mathématiques Pures et Appliquées* 42 (1963), 21–27.

[124] E. T. Parker, Edge-coloring numbers of some regular graphs. *Proceedings of the American Mathematical Society* 37 (1973), 423–424.

[125] J. Petersen, Die Theorie der regulären Graphs. *Acta Mathematica* 15 (1891), 193–220.

[126] L. Petrenyuk and A. Petrenyuk, Intersection of perfect one-factorizations of complete graphs. *Cybernetics* 16 (1980), 6–9.

[127] J. Pila, Connected regular graphs without one-factors. *Ars Combinatoria* 18 (1983), 161–172.

[128] T. Pisanski, J. Shawe-Taylor and B. Mohar, 1-factorization of the composition of regular graphs. *Publications de l'Institut Mathématique* (Nouvelle Série) 33 (47) (1983), 193–196.

[129] L. Pósa, A theorem concerning Hamilton lines. *Magyar Tudományos Akadámia Matematikai Kutató Intézetének Közleményei* 7 (1962), 225–226.

[130] R. Rees and W. D. Wallis, The spectrum of maximal sets of one-factors. *Discrete Mathematics* 97 (1991), 357–369.

[131] D. F. Robinson, Constructing an annual round-robin tournament played on neutral grounds. *Mathematical Chronicle* 10 (1981), 73–82.

[132] A. Rosa and W. D. Wallis, Premature sets of 1-factors, or, how not to schedule round-robin tournaments. *Discrete Applied Mathematics* 4 (1982), 291–297.

[133] K. G. Russell, Balancing carry-over effects in round robin tournaments. *Biometrika* 67 (198), 127–131.

[134] T. Schönberger, Ein Beweis des Petersenschen Graphensatzes. *Acta Universitatis Szegediensis. Acta Scientiarum Mathematicarum* 7 (1934), 51–57.

[135] E. Seah, Perfect one-factorizations of the complete graph – a survey. *Bulletin of the Institute of Combinatorics and its Applications* 1 (1991), 59–70.

[136] E. Seah and D. R. Stinson, Some perfect one-factorizations for K_{14}. *Annals of Discrete Mathematics* 34 (1987), 419–436.

[137] E. Seah and D. R. Stinson, On the enumeration of one-factorizations of the complete graph containing prescribed automorphism groups. *Mathematics of Computation* 50 (1988), 607–618.

[138] E. Seah and D. R. Stinson, A perfect one-factorization for K_{36}. *Discrete Mathematics* 70 (1988), 199–202.

[139] E. Seah and D. R. Stinson, A perfect one-factorization for K_{40}. *Congressus Numerantium* 68 (1989), 211-214.

[140] R. G. Stanton and I. P. Goulden, Graph factorizations, general triple systems and cyclic triple systems. *Aequationes Mathematicae* 22 (1981), 1–28.

[141] G. Stern and H. Lenz, Steiner triple systems with given subspaces: Another proof of the Doyen–Wilson theorem. *Bolletino della Unione Matematica Italiana* A, Serie V, 17 (1980), 109–114.

[142] D. R. Stinson, Some constructions for frames, Room squares, and subsquares. *Ars Combinatoria* 12 (1981), 229–268.

[143] D. R. Stinson, Some results concerning frames, Room squares, and subsquares. *Journal of the Australian Mathematical Society* 31A (1981), 53–65.

[144] D. R. Stinson, A short proof of the nonexistence of a pair of orthogonal Latin squares of order six. *Journal of Combinatorial Theory* 36A (1984), 373–376.

[145] T. H. Straley, Scheduling designs for a league tournament. *Ars Combinatoria* 15 (1983), 193–200.

[146] A. P. Street and W. D. Wallis, *Combinatorics: A First Course.* (Charles Babbage Research Centre, Winnipeg, 1983.)

[147] G. Tarry, Le problème des 36 officiers. *Comptes Rendus de L'Association Française pour L'Avancement de Science* 1 (1900), 122–123; 2 (1901), 170–203.

[148] L. Teirlinck, On the use of pairwise balanced designs and closure spaces in the construction of structures of degree at least 3. *Le Matematiche* 65 (1990), 197–218.

[149] L. Teirlinck, Large sets of disjoint designs and related structures. In *Contemporary Design Theory* (Wiley, New York, 1992), 561–592.

[150] W. T. Tutte, The factorizations of linear graphs. *Journal of the London Mathematical Society* 22 (1947), 459–474.

[151] W. T. Tutte, On the problem of decomposing a graph into n connected factors. *Journal of the London Mathematical Society* 36 (1961), 221–230.

[152] V. G. Vizing, On an estimate of the chromatic class of a p-graph [Russian]. *Discretyni Analiz* 3 (1964), 25–30.

[153] D. Wagner, On the perfect one-factorization conjecture. *Discrete Mathematics* 104 (1992), 211–215.

[154] W. D. Wallis, On one-factorizations of complete graphs. *Journal of the Australian Mathematical Society* 16 (1973), 167–171.

[155] W. D. Wallis, Solution of the Room square existence problem. *Journal of Combinatorial Theory* 17A (1974), 379–383.

[156] W. D. Wallis, The smallest regular graphs without one-factors. *Ars Combinatoria* 11 (1981), 295–300.

[157] W. D. Wallis, One-factorizations of wreath products. *Combinatorial Mathematics VIII* (*Lecture Notes in Mathematics* 884), Springer-Verlag (1981), 337–345.

[158] W. D. Wallis, A one-factorization of a cartesian product. *Utilitas Mathematica* 20 (1981), 21–25.

[159] W. D. Wallis, The problem of the hospitable golfers. *Ars Combinatoria* 15 (1983), 149–152.

[160] W. D. Wallis, A tournament problem. *Journal of the Australian Mathematical Society* 24B (1983), 289–291.

[161] W. D. Wallis, One-factorizations of graphs: Tournament applications. *College Mathematics Journal* 18 (1987), 116–123.

[162] W. D. Wallis, *Combinatorial Designs.* (Marcel Dekker, New York, 1988.)

[163] W. D. Wallis, A class of premature sets of one-factors. *Annals of the New York Academy of Sciences* 555 (1989), 425–428.

[164] W. D. Wallis, The lengths of trains. *Congressus Numerantium* 77 (1990), 199–204.

[165] W. D. Wallis, Staircase factorizations. *Gazette of the Australian Mathematical Society* 18 (1991), 1–3.

[166] W. D. Wallis, One-factorizations of complete graphs. In *Contemporary Design Theory* (Wiley, New York, 1992), 593–631.

[167] W. D. Wallis, A. P. Street and J. R. S. Wallis, *Combinatorics: Room Squares, Sum-free Sets, Hadamard Matrices* (Lecture Notes in Mathematics 292). (Springer-Verlag, Heidelberg, 1972.)

[168] W. D. Wallis and Z. Wang, On one-factorizations of cartesian products. *Congressus Numerantium* 49 (1985), 237–245.

[169] W. D. Wallis and Z. Wang, Some further results on one-factorizations of cartesian products. *Journal of Combinatorial Mathematics and Combinatorial Computing* 1 (1987), 221–234.

[170] J. Wang, On factorizations of graphs [Chinese]. *Chinese Quarterly Journal of Mathematics* 3 (1988), 37–51.

[171] D. de Werra, Scheduling in sports. In *Studies on Graphs and Discrete Programming* (North-Holland, Amsterdam, 1981), 381–395.

[172] D. de Werra, On the multiplication of divisions: The use of graphs for sports scheduling. *Networks* 15 (1985), 125–136.

[173] D. de Werra, L. Jacot-Descombes and P. Masson, A constrained sports scheduling problem. *Discrete Applied Mathematics* 26 (1990), 41–49.

[174] H. S. White, Triple systems as transformations and their paths along triads. *Transactions of the American Mathematical Society* 14 (1913), 6–13.

[175] H. P. Yap, *Some Topics in Graph Theory*. (Cambridge University Press, Cambridge, 1986.)

[176] C.-Q. Zhang, On a theorem of Hilton. *Ars Combinatoria* 27 (1989), 66–68.

[177] C.-Q. Zhang and Y.-J. Zhu, Factorizations of regular graphs. *Journal of Combinatorial Theory* 56B (1992), 74–89.

INDEX